9-Object-Model FMEA

One thing I've always found remarkable, and one that seems to connect many of the great thinkers, scientists, and geniuses throughout history, is the fact that they never say anything about their exceptional intellectual abilities.

Instead, they limit themselves to telling humanity that they feel constantly animated by an inextinguishable thirst for knowledge, an unquenchable curiosity, and a ceaseless need to ask themselves questions.

Pier Luigi Belcaro

9-Object-Model FMEA

A Conceptual Model for Creating Technical Risk Analyses

Springer

Pier Luigi Belcaro
Battenberg (Eder), Germany

ISBN 978-3-658-50831-9 ISBN 978-3-658-50832-6 (eBook)
https://doi.org/10.1007/978-3-658-50832-6

This book is a translation of the original German edition "9-Object-Model FMEA" by Pier Luigi Belcaro, published by Springer Fachmedien Wiesbaden GmbH in 2025. The translation was done with the help of an artificial intelligence machine translation tool. A subsequent human revision was done primarily in terms of content, so that the book will read stylistically differently from a conventional translation. Springer Nature works continuously to further the development of tools for the production of books and on the related technologies to support the authors.

Translation from the German language edition: "9-Object-Model FMEA" by Pier Luigi Belcaro, © Der/die Herausgeber bzw. der/die Autor(en), exklusiv lizenziert an Springer Fachmedien Wiesbaden GmbH, ein Teil von Springer Nature 2025. Published by Springer Fachmedien Wiesbaden. All Rights Reserved.

This Springer imprint is published by the registered company Springer Fachmedien Wiesbaden GmbH, part of Springer Nature.
The registered company address is: Abraham-Lincoln-Str. 46, 65189 Wiesbaden, Germany

I dedicate this second book wholeheartedly to the memory of Don Lorenzo Milani, an extraordinary man and unparalleled teacher.

To him and to all great men like him, who with infinite generosity and indomitable courage, have always fought to promote a model of transmission and exchange of knowledge aimed at an authentic, genuine progress of humanity.

A model that, precisely because of the highest goals it proposes, cannot and must not be separated from a profound devotion to the highest and noblest moral principles: humility, a spirit of welcome, universal fraternity.

And the sincere will to leave no one behind.

Preface

This second book appears approximately six years after my first short essay on FMEA:

"FMEA: The Technical Industrial Risk of Product and Process"
was published in January 2018 under the direction of SANDIT Libri.

At that time, my primary intention was to provide a brief, informative overview of FMEA techniques, from the specific perspective of the German VDA standard and with the specific digital support of the German software APIS-IQ.

This first book already contained some characteristic features, albeit still in their infancy, that play a key role in the present work. I refer in particular to the *conceptual model*, which I personally consider the basis for all treatment of FMEA.

The German VDA standard, when studied thoroughly and understood in all its powerful implications, already provides excellent support for further methodological/operational developments. What I have done in this sense as original and perhaps as a "plus" (and this second book is metaphorically the stage for this second act of the opera) is that I have designed a complete and structured exposition on this key to interpreting the FMEA topic, an *exposition based on the centrality of a conceptual model with nine objects.*

In this book, I therefore introduce and deepen the topic of the "*Conceptual Model*", describing and explaining what the nine objects at its core are, the interactions these nine objects enter into, and the implications that can arise from them, both at the primary methodological/analytical level and at the tactical/operational level as well.

This work also continues to benefit from the *support of the APIS-IQ software*, which even takes on crucial characteristics here through the application of some of its particularly useful "advanced options."

Based on the robust foundations of the conceptual model with nine objects, you will find *two complete structured procedures* here: one for *Design FMEA* and one for *Process FMEA*. Everything that initially had absolutely relevant significance on the methodological/formal side later regains this significance on the operational side by providing a complete FMEA process. And since the general approach here is based entirely on the principle of "scientific rigor," a treatment (also structured) of the *logical junction between a D-FMEA and its corresponding P-FMEA* could not be missing.

Finally, you'll find the *mathematical model* I presented in my first book here. You'll find it for a very specific reason: In accordance with the systemic/modelistic approach and its logical implications, the mathematical model is intended to serve as a stimulus for a further possible development of the entire "FMEA model".

Battenberg (Eder), Germany Pier Luigi Belcaro

Acknowledgments The author would like to express his particular gratitude to *APIS Informationstechnologien GmbH* for their kind permission to publish images, some of which were derived from original APIS-IQ images or, in any case, from images associated with the APIS-IQ software.

Personally, I consider the APIS-IQ software to be *an excellent tool*, without whose valuable support *my daily work as an FMEA Coordinator would simply be inconceivable.*

Competing Interests The author has no competing interests to declare that are relevant to the content of this manuscript.

Contents

Introduction and Goals of the Book

1

Abstract

This chapter provides a concise overview of three key points: what FMEA is, its purpose, and why it is so widely used. Following this, the book's three goals are outlined:

- A proposal for an FMEA methodology (and resulting operational procedures) specifically based on a nine-object conceptual model;
- A comprehensive overview of the whole FMEA cycle: design, production process, and product/process interface;
- Ideas for potential further developments of FMEA. This chapter introduces a mathematical model that serves as an alternative to traditional rating tables in the FMEA assessment phase for risk parameters Occurrence and Detection.

1.1 What's FMEA?

FMEA (Failure Mode and Effects Analysis) is an established system (a technique, a methodology) that is extensively used in the field of preventive quality in many industries.

This system focuses on the analysis of products and the corresponding production processes. The goal is to identify potential problems (errors/failures) early in a product's life cycle and to analyze them thoroughly in order to define and implement the most appropriate actions while keeping the costs of these actions as low as possible.

All this translates into ***high product/process quality*** and, at the same time, ***savings*** (some of which are very significant) ***in the corresponding industrial costs***.

P. L. Belcaro, *9-Object-Model FMEA*, https://doi.org/10.1007/978-3-658-50832-6_1

1.1.1 FMEA: Three Key Questions

What is FMEA?

- ***FMEA is a technique***, a methodology—or systematic, as we said above—***inductive and analytical***, which operationally, at least from the German perspective of the VDA standard, consists of ***a well-defined sequence of "logical steps"***.

What is it for?

- The purpose of FMEA is the ***structural, functional and dysfunctional analysis*** of a product or an industrial production process, carried out with the main goal of ***preventively identifying all possible problems (errors/failures) and ranking them in order of increasing priority***, so that targeted actions can then be defined and implemented aimed at the most effective management (technical and economic) of the problems.

Why is it widely used?

- FMEA is primarily used ***for very specific and absolutely relevant economic reasons***. In modern industrial contexts, it can indeed be easily demonstrated that adhering to the "ASAP" principle—As Soon As Possible (implied: within a product's life cycle)—with regard to actions brings decisive economic benefits (see Fig. 1.1).

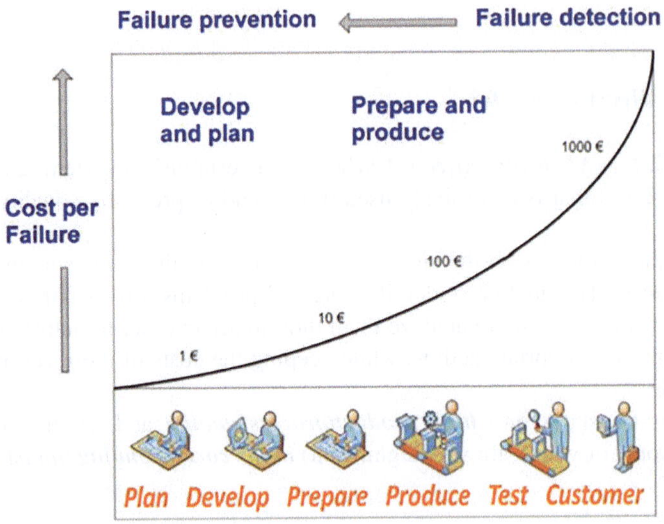

Fig. 1.1 "Powers of 10" curve

1.2 Goals of the Book

This book has the following three main goals:

- To propose an FMEA systematic and operational procedures derived from it, based on a very precise ***CONCEPTUAL MODEL WITH 9 OBJECTS***, so that they not only fully comply with the standard (VDA), but are also highly structured and strictly scientifically sound.
- To cover the entire ***FMEA logic cycle*** by describing, explaining, and commenting on complete didactic examples of D-FMEA and P-FMEA. The essential topic of the "***Product/Process interface***" (the transition from D-FMEA to P-FMEA) is here also addressed specifically and in detail.
- To provide ***suggestions for possible "advanced development directions"***. In this case, I am referring in particular to the basic mathematical model I developed in 2015 to identify the risk parameters Occurrence and Detection in the evaluation phase of an FMEA. In this case, I deliberately do not use evaluation tables, but rather a mathematical model. All this with the idea of ***reducing the subjectivity*** of the assessments implicit in the tabular definitions as much as possible.

Genesis, History and Standards

2

Abstract

This chapter addresses the origins of FMEA. The international standards (SAE/AIAG, VDA, AIAG/VDA) that govern FMEA are then described, emphasizing a key difference between the American and German standards: the systemic model (structure/functions/failures). The VDA standard is explained in detail at the end. The five steps of the standard, the activities involved in each step, and the systemic model with its horizontal layers of analysis, vertical development levels, systemic hierarchies, and logical relationships are then examined as further points of focus.

2.1 Origin and History of FMEA

The FMEA technique was originally developed and used by NASA in the early 1960s to ensure the overall quality of the Apollo projects.

After its initial use in space technology, FMEA was introduced in the automotive industry by the Ford Motor Company for preventive quality assurance starting in the mid-1970s.

A standard for FMEA has existed in Germany since 1980. Further development of the method continues primarily in the automotive sector.

However, ***FMEA can be used regardless of the product or sector***. Its main application is currently the reliability analysis of new developments and product changes in safety-relevant systems, as is the case today, for example, in the automotive industry, power plant technology, medical applications, and the aerospace industry. Thanks to its success in these areas, the method is also widely used in other contexts, such as service quality assurance and project management.

P. L. Belcaro, *9-Object-Model FMEA*, https://doi.org/10.1007/978-3-658-50832-6_2

2.2 International Standards

Historically, ***there are two international standards*** that rule and structure the FMEA system:

- The ***American SAE/AIAG standard***
- The ***German VDA standard***.

Since June 2019, a so-called

- ***"Harmonized" AIAG/VDA standard***

has also come into force.

This is the result of the work of a joint technical commission (SAE and VDA experts) whose goal was to unify the two standards—as far as possible, given the obvious differences in their methodological principles—which have always existed independently of each other.

Personally, however, I believe that the harmonized standard has actually added or improved nothing significant compared to a complete and well-structured application of the German standard. For this reason, from now on, I will only refer to the two original standards.

Every risk analysis (or FMEA) is therefore conducted in accordance with one of these two standards. The difference between the two standards is very significant. The main reason is that the two standards are based on very different methodological assumptions.

According to the American standard, the focus is on investigating failures and the corresponding analysis of actions, so that all that matters for potential failures is their identification/presentation in a sequence.

"Failure Modes"—"Failure Causes"—"Failure Consequences."

To conduct a risk analysis according to the American standard, MS Excel software (spreadsheets) is typically used.

The German standard, on the other hand, considers a well-defined "working model" as the central issue, which should be deeply understood and appropriately applied ***before*** failure analysis begins. In this case, it is ***a "systemic" model*** (i.e., a "System") because ***(logical) levels and corresponding systemic hierarchies*** are clearly recognizable in this model.

Working according to the German standard absolutely requires the use of specific software, as the objects belonging to the various levels of the model and the underlying logical and hierarchical relationships could never be managed using a simple Excel spreadsheet. There is an absolute need for an ***RDBMS (Relational Database Management System)***, or, to be more precise, for ***"RDBMS-based" software.***

System Analysis	01	***Structure Analysis***
	02	***Function Analysis***
	03	***Failure Analysis***
Risk Analysis	04	***Action Analysis and Risk Evaluation***
	05	***Optimization***

Fig. 2.1 The VDA 5-step method

The ***IQ-FMEA software (or its more powerful version IQ-RM) from the German software house "APIS Informationstechnologien GmbH"*** has, in my opinion, been the best choice in the field of VDA risk analyses for many years.

2.3 The VDA 5-Step Method

A VDA-compliant FMEA includes the following 5 steps (see Fig. 2.1).

Please note: ***Steps 1 and 2 are performed interdependently***, and this is ***the only exception*** in the method. Below is an image (see Fig. 2.2) describing the activities involved in each step.

What we said above about the "system" according to the German standard and the "systemic working model" can now be better understood by taking a closer look at the three logical (horizontal) levels (levels of analysis) of the model (see Fig. 2.3).

PLEASE NOTE: The VDA 5-Step Method, as described in the official VDA documentation, proposes specific terminology for the following objects:

- The 6 structural objects (3 for the D-FMEA and 3 for the P-FMEA)
- The 3 functional objects of the D-FMEA
- The 3 failure objects (they are identical for both D-FMEA and P-FMEA).

However, the terminology of the 3 objects related to the functional layer of the P-FMEA is my personal choice.

At this point, it can be clearly understood that ***A SYSTEM CONSISTING OF EXACTLY 9 OBJECTS IS (NECESSARY AND) SUFFICIENT TO PERFORM AN FMEA ACCORDING TO THE GERMAN STANDARD***.

We require exactly ***three logical layers (hierarchically ordered analysis layers)***, with exactly ***three required objects per layer***.

Hierarchical relationships can be viewed vertically (***between analysis layers***): What is determined and performed at the second layer is strictly and directly dependent on what was previously determined and performed at the first layer. This also applies, of course, when we consider the third layer in comparison to the second. The functions simply wouldn't make any sense if we hadn't first established a structure and organized it logically. Furthermore, a failure analysis would

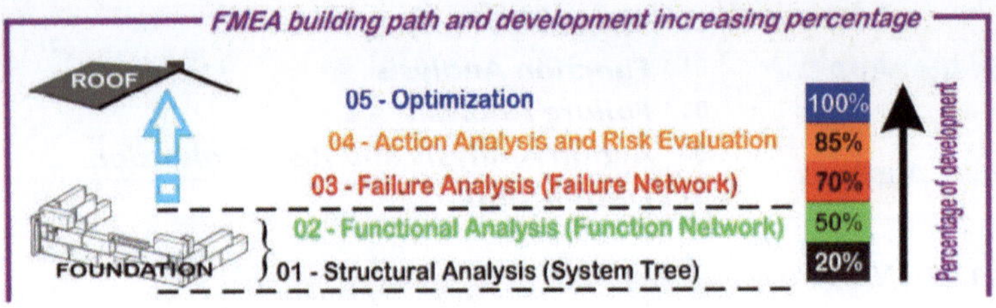

01 - Structure Analysis
- Identify all involved system (physical) elements
- Create a structure tree
- Conduct a selection analysis

02 - Function Analysis
- Assign functions to system elements
- Logically link functions
- Create a functional network

03 - Failure Analysis
- Derive failures from the functions
- Logically link failures
- Create a failure network

04 - Action Analysis and Risk Evaluation
- Add prevention and detection actions to the form
- Assign responsible persons and deadlines to the actions
- Evaluate/Estimate Severity, Occurrence and Detection

05 - Optimization
- Reassess after implementation of the actions
- If results are unsatisfactory (according to a defined optimization criterium), set a new action status
- Back to Step 4

Fig. 2.2 The VDA 5-step method (individual activities)

be pointless if we hadn't previously defined any function and the logical sequence of them.

Finally, horizontally (between objects at the same analysis layer), ***logical relationships (of the "IF – THEN" type, i.e., cause/effect relationships)*** are recognizable. In the model, this applies to the last two analysis layers: the function layer and the failure layer (see Fig. 2.4).

As for the technical content of each step, we will see everything in detail in Sects. 4.3, 4.4 and 4.5 (D- and P-FMEA, operational procedures).

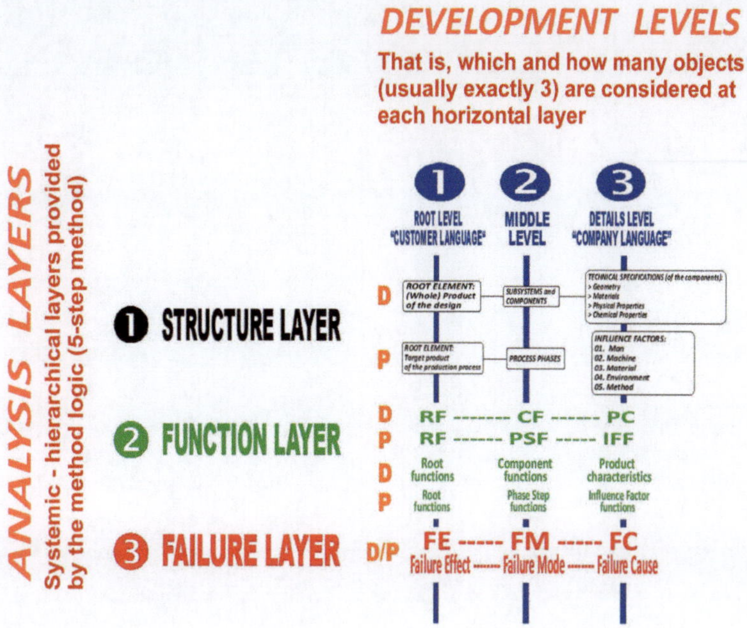

Fig. 2.3 Working model (layers and levels)

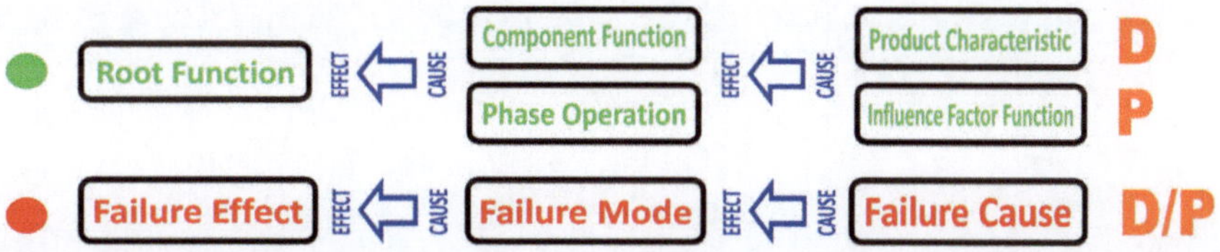

Fig. 2.4 Logical relationships

FMEA: The Conceptual Model

3

Abstract

This chapter introduces the conceptual model that forms the core of the entire book, explaining the role of the model's nine components, the layers of analysis, the development levels, the sorting, the systemic hierarchies, and the logical relationships. Finally, a "dynamic" perspective of the model—that is, the logical/chronological sequence of the analysis layers during an FMEA—is presented.

3.1 The Conceptual Model: What and How

The presentation of the VDA approach I will make here could correctly be considered a "***systemic composition***". What do I mean by that? I mean that ***I will deliberately consider certain "objects", taking into account some of their defined "properties"***.

A more detailed explanation follows.

A ***risk analysis (FMEA), as presented and treated by the German VDA standard***, is nothing other than ***a "system"*** (a "***systemic representation***", that is, ***a structured composition of objects*** in which ***the layers of analysis (horizontal)***, as well as ***the development levels (vertical)***, and the ***hierarchies (systemic hierarchical relationships between the analysis layers)*** can be represented as follows (see Fig. 3.1).

In other words, we could say that ***a system consisting of (only) 9 objects would be perfectly sufficient to conduct an FMEA (optimally)***. We need ***three layers of analysis plus three development levels, whose intersections identify exactly 9 objects***.

In the figure above (see Fig. 3.1), a distinction is made between ***design objects (D)*** and ***process objects (P)***.

P. L. Belcaro, *9-Object-Model FMEA*, https://doi.org/10.1007/978-3-658-50832-6_3

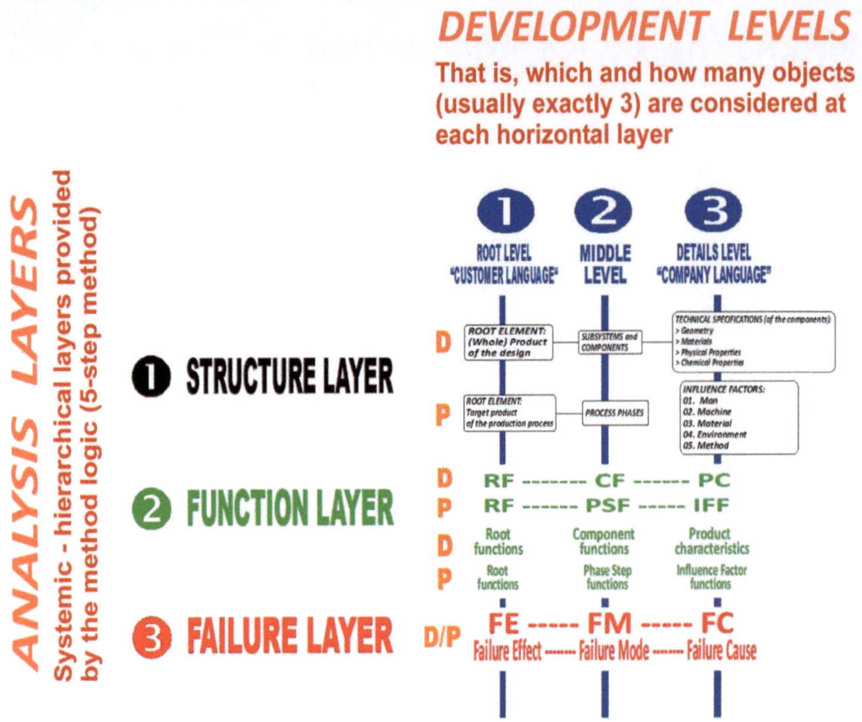

Fig. 3.1 Systemic representation of FMEA

If we wanted to examine this representation in more detail, we could look at the following figure (see Fig. 3.2).

where I added the terms "***hierarchy***" (or "***super- and subordination***") and "***logical relationship***".

We could also say that ***hierarchies represent a kind of*** "***vertical order***" (i.e., an order between horizontal layers) and ***logical relationships represent a kind of*** "***horizontal order***" (i.e., an order between vertical levels).

In summary:

- ***3 Analysis layers (horizontal)***
  - Structural layer
  - Functional layer
  - Malfunction layer (or "Failure layer")
- ***3 Development levels (vertical)***
  - Root level (the so called "customer language")
  - Middle level
  - Details level (the so called "company language")
- ***An H-ordering (horizontal order***), that is, an arrangement between the vertical levels of development: the logical cause-effect relationships

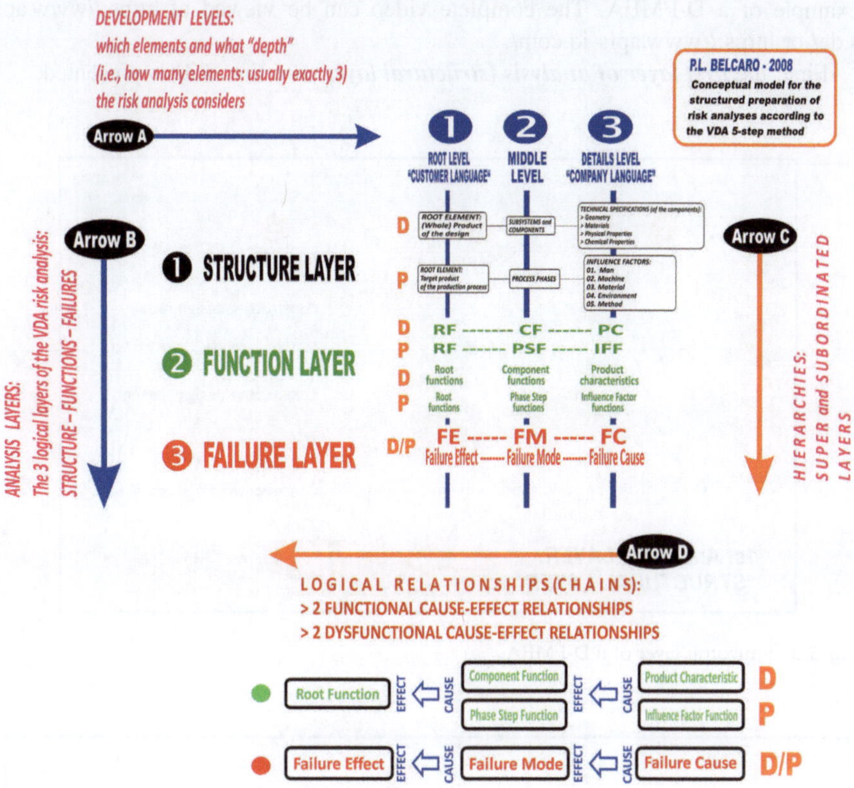

Fig. 3.2 The systemic conceptual model

- ***A V-ordering (vertical order)***, i.e. an arrangement between the horizontal layers: the systemic hierarchies (superior and subordinate elements)
- ***9 objects (at the intersections of the three layers of analysis with the three levels of development)***, differentiated into D (design) and P (process), as shown in the image above.

are ***exactly all the*** "***ingredients***" that we actually need to create our "***structured and efficiency-maximizing representation***" of an FMEA.

3.2 Further Considerations on the Model: A Dynamic Development

Here I would just like to suggest a few illustrations that I believe could be very useful for further understanding everything I've described above.

These are still images that I isolated from a video by ***APIS Informationstechnologien GmbH***. This video introduces the ***APIS-IQ (v. 7.0) software*** using an example of a D-FMEA. The complete video can be viewed at https://www.apis.de/ or https://www.apis-iq.com/.

First, the ***first layer of analysis (structural layer: see*** Fig. 3.3***)*** is presented.

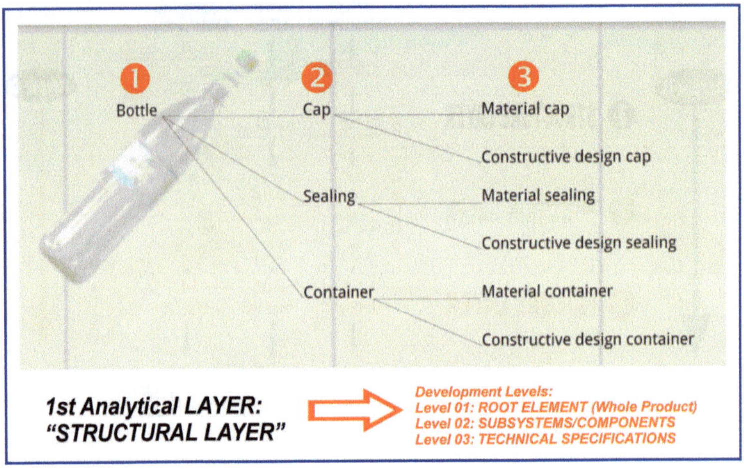

Fig. 3.3 Structural layer of a D-FMEA

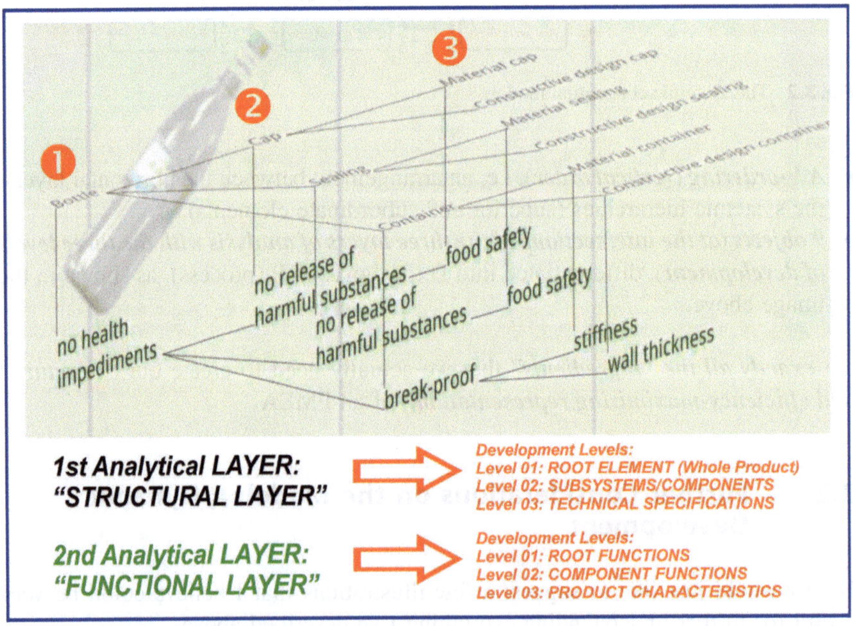

Fig. 3.4 Functional layer of a D-FMEA

Then comes the ***second layer of analysis (functional layer: see*** Fig. 3.4*)*, which can also be considered as the “first subordinate level”.

Finally, the ***third and final layer of analysis (Failure layer: see*** Fig. 3.5*)* follows, which can also be called the “second subordinate level”.

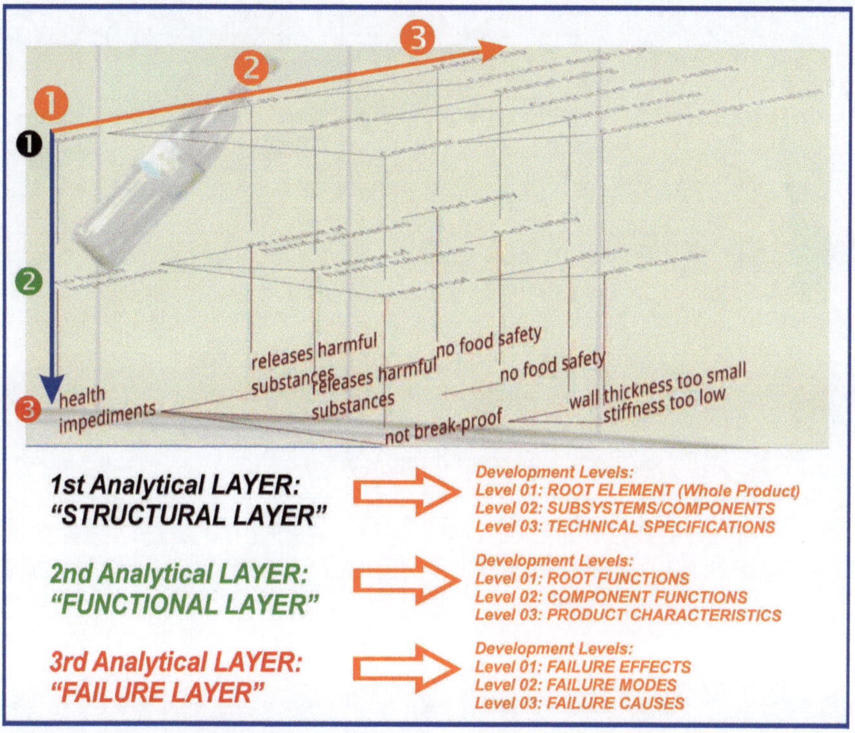

Fig. 3.5 Failure layer of a D-FMEA

FMEA: The Operational-Process Aspect

4

Abstract

This chapter explains how the operational procedures (D-FMEA and P-FMEA) were developed based on the conceptual model, describing the logical blocks that specifically comprise the D-FMEA and P-FMEA procedures, as well as the common blocks (D- and P-FMEA).

4.1 Introduction: Specific Technical Terminology of FMEA Objects

Before we begin the section on operational procedures, we will define the ***technical terminology*** that we will use systematically in this discussion.

The following technical terms (see Fig. 4.1) refer to the ***objects of an FMEA***, i.e., the ***objects of the conceptual model***, as we saw above.

4.2 Composition of the Operational Procedures: Specific and Common Logical Blocks

The two ***procedures (D-FMEA and P-FMEA)*** presented here consist of four "logical blocks," each containing macro and micro activities that relate to specific steps of the VDA method or to its aggregates.

The first two logical blocks:

- ***BLOCK 00***: PREPARATORY ACTIVITIES, development of the structural tree and basic functional frame (VDA Method: Step 1 and part of Step 2)
- ***BLOCK 01***: SYSTEM ANALYSIS (VDA method: Step 2 and Step 3 completed)

P. L. Belcaro, *9-Object-Model FMEA*, https://doi.org/10.1007/978-3-658-50832-6_4

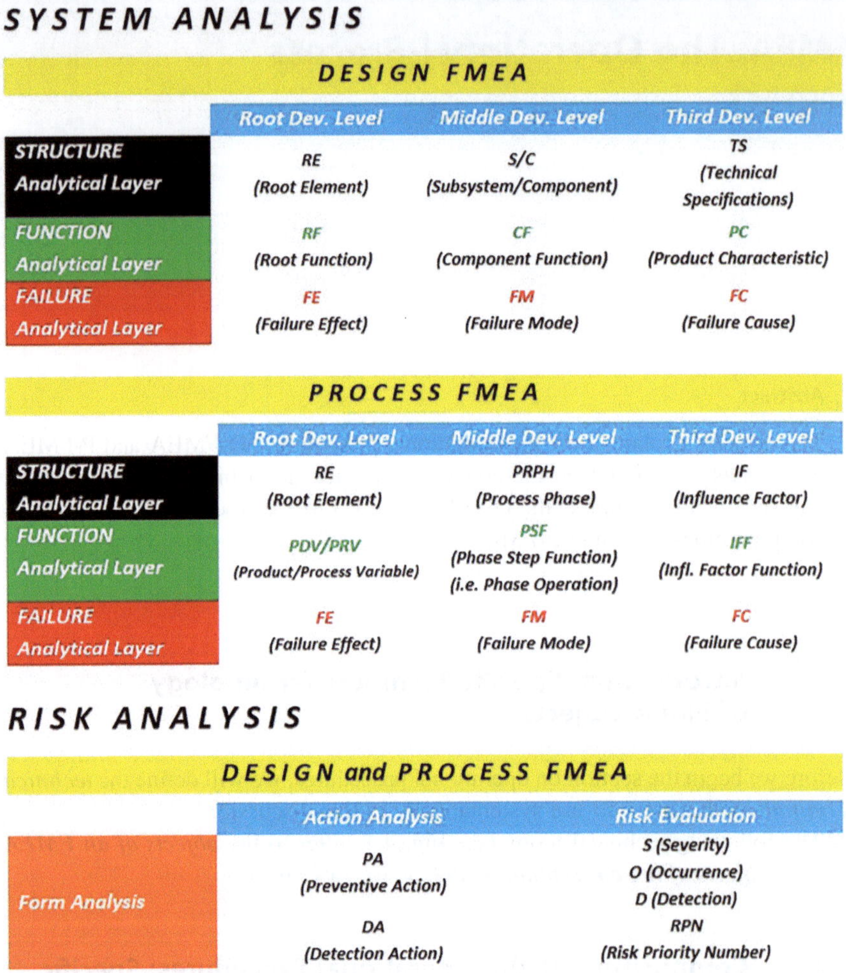

Fig. 4.1 Technical terminology of FMEA objects

are ***specific to the subject of the FMEA (product or production process)***. This means that the described activities, even if they share a common logical thread, are structured differently depending on whether it is a D-FMEA or a P-FMEA.

However, the last two blocks:

- ***BLOCK 02***: RISK ANALYSIS, Part 1 (VDA Method: Step 4)
- ***BLOCK 03***: RISK ANALYSIS, Part 2 (VDA Method: Step 5)

include elements and activities that are largely interchangeable between D-FMEA and P-FMEA, although the thematic diversity in concrete examples is clearly

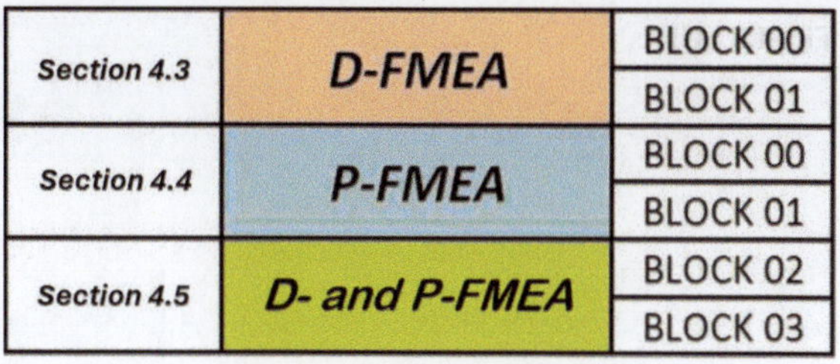

Fig. 4.2 Procedures and logical blocks

linked to the different nature of the object being addressed (design product or production process).

Based on the circumstances described above, the procedures are presented as follows (see Fig. 4.2).

4.3 D-FMEA Operational Procedure

4.3.1 Logical Block 00: Preparatory Activities, Development of the Structural Tree and of the Essential Functional Backbone (VDA Method: Step 1 and part of Step 2)

4.3.1.1 Input Data (RF List with Component Propagations)

The starting point for a design FMEA is the ***input data***, i.e. the data that the FMEA coordinator usually receives from the colleagues in the design department who created the 3D CAD data and the technical drawings of the product.

Data Format (Txt Text File)

The data regarding ***structure*** and ***main functional frame*** should best be provided to the FMEA coordinator through simple text files (files with the extension TXT) in order to be operationally useful for him.

Text File Layout

The relevant TXT files should be compiled in a format suitable for single lines (SLL: Single Line List). Here we see an example of a hypothetical product "glass" (made of polymer composite), an example introduced in my first book (see Fig. 4.3).

The TXT file in question (see Fig. 4.4) could be, for example, the following.

Fig. 4.3 Glass

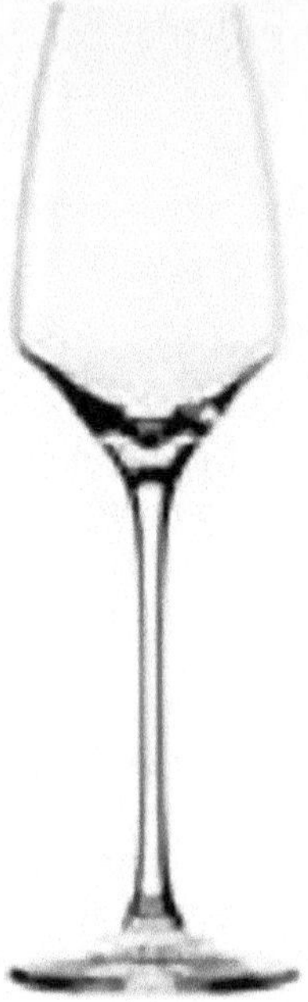

4.3.1.2 Data Import (into APIS-IQ)

Root Node RE (Root Element) and Root Functions (RF)

Components (S/C) and Component Functions (CF)

Based on a TXT file with the above-mentioned properties, the coordinator prepares the APIS-IQ D-FMEA file appropriately ***before opening the first session***.

By using APIS-IQ's very useful advanced options for external data import, the final result (see Fig. 4.5) will look something like this:

As we can see from the figure above, ***almost the entire structural tree and functional frame are already in place at this point***, with the exception of the third-level objects: TS (structural level) and PC (functional level). All of this represents ***a major advantage in terms of the operational efficiency of the FMEA analysis***.

The relevant point here, however, is to ***carefully avoid the analysis not being prepared/automated as much as possible***. Otherwise, one risks not only obvious operational inefficiencies but also boring or giving a bad impression to technical department colleagues by involving them in initial meetings that are actually unnecessary.

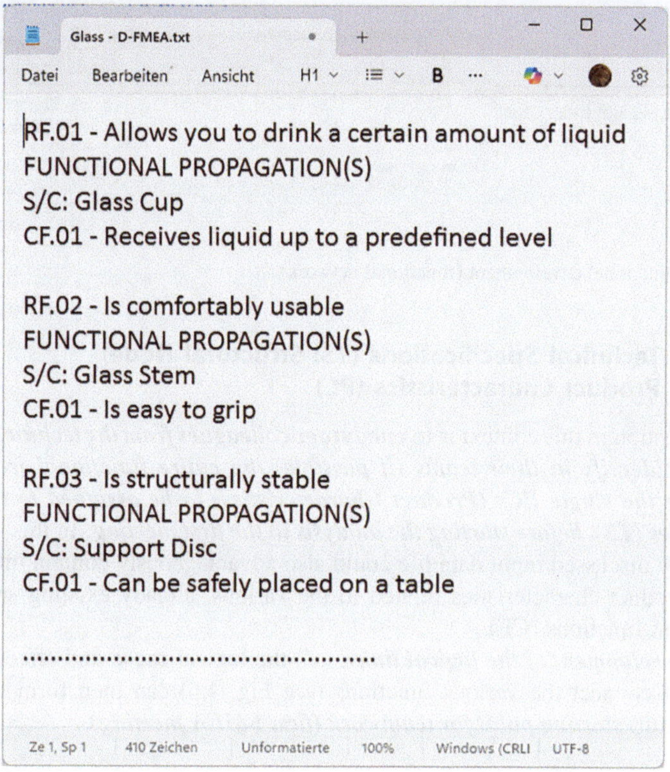

Fig. 4.4 TXT file "Glass"

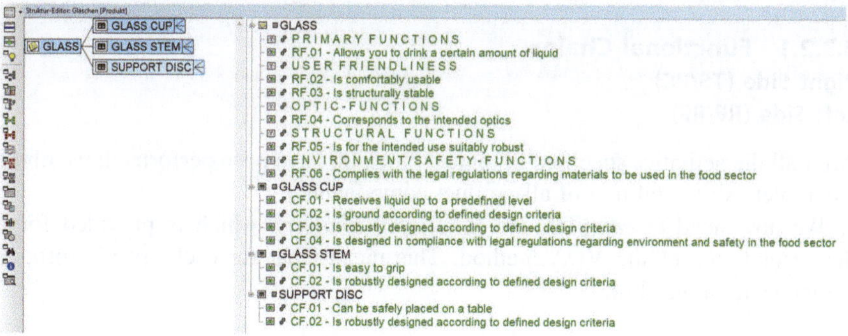

Fig. 4.5 Structure and functional framework after import

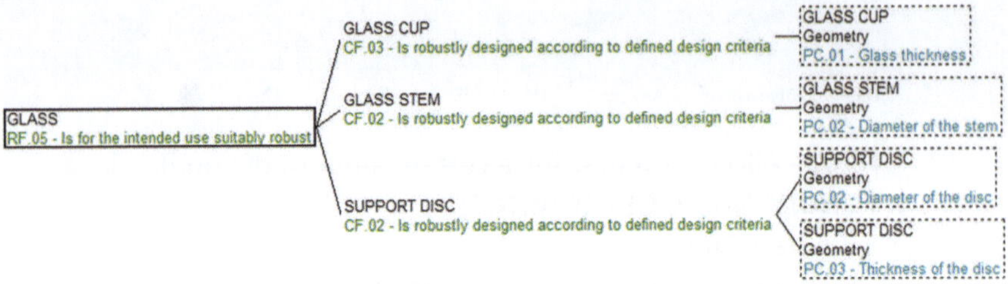

Fig. 4.6 Functional development (functional network)

4.3.1.3 Technical Specifications (TS: Structural Node)
4.3.1.4 Product Characteristics (PC)

My suggestion in this context is to ***encourage colleagues from the technical departments to identify in their teams (if possible) the entire functional framework, including the single PCs (Product Characteristics) to be assigned to the structural nodes (TS), before starting the analysis in the first meeting***. In this sense, the previously discussed input data file could also advantageously contain information on the product characteristics related to the various, already existing specialized component functions (CF).

The ***development of the logical links***, i.e., the logical cause-and-effect relationships that connect the various functions (see Fig. 4.6) can then form the actual (and fruitful) ***starting point for teamwork (first FMEA meeting)***.

4.3.2 LOGICAL BLOCK 01: SYSTEM ANALYSIS (VDA Method: Step 2 and Complete Step 3)

4.3.2.1 Functional Chains
Right Side (TS/PC)
Left Side (RE/RF)

After all the activities specified in logical block 00 have been performed, we obtain a complete structural tree of all product components.

We now need to ***complete the functional analysis***, which is provided for in the second step of the VDA method. This means that for each specific product component, we need to:

- Create the structural node "Technical Specifications" (TS) and identify the relevant Product Characteristics (PC)
- Link the product characteristics to the component functions (right side of the function chain),
- Identify all applicable/relevant root functions listed under the root element

- Link the root functions to the component functions (left side of the function chain)

An important clarification: When we speak of "***Technical Specifications***" in a D-FMEA, we generally refer to ***3 well-defined thematic classes***:

- Geometry
- Materials
- Physical and chemical properties

In any case, ***a Technical Specification always refers to measurable quantities***, i.e., ***numerical quantities with their own measurement unit***. An example to better illustrate the concept:

Geometry	• Linear dimensions • Diameters • Radii of curvature • Angles • Curves (described by analytic functions) • Surfaces (described by analytic functions) • Volumes (described by 3D analytic functions) • Etc
Materials	• Technical material definitions (e.g. hardness, porosity, hygroscopicity, etc.)
Physical and chemical properties	• Electrical resistance • Pressure • Performance • Corrosiveness • Flammability • Reaction to acids • Etc

Let us now give an ***example of the logical functional development of a component function***. Suppose we continue to analyze the example of the small glass mentioned above and consider the component "support disc" with the following CF:

- **CF.01—Can be safely placed on a table**

First, we should ask ourselves some technical questions:

- ***How many and which TS*** should be correctly considered here?
- Even better: ***which TS and which PC*** could best explain and justify (i.e., according to the strictest possible cause-effect logic) the CF under investigation?

In this case, the following might be an appropriate technical response:

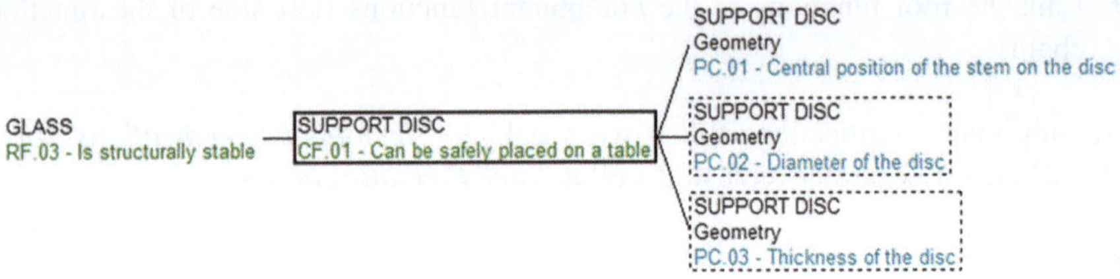

Fig. 4.7 Logical function chain (complete)

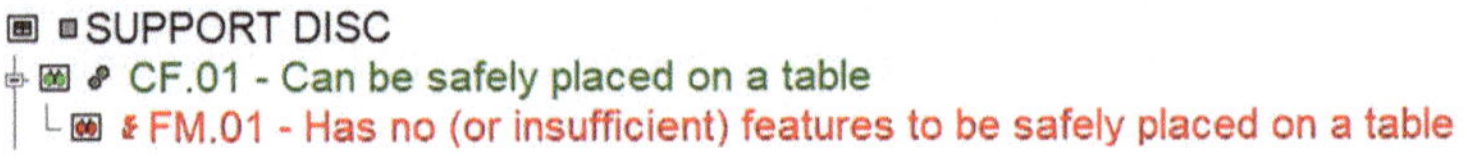

Fig. 4.8 Derivation of a FM from a CF

1. TS SUPPORT DISC/PC.01: Center position of the stem on the disc (GEOM.)
2. TS SUPPORT DISC/PC.02: Disc diameter (GEOM.)
3. TS SUPPORT DISC/PC.03: Disc thickness (GEOM.)

Now we only need the left side of the chain, i.e., the second (and final) logical cause-effect relationship. To achieve this result, we proceed in the same way as before. We first identify the relevant root functions under the root element and concatenate them with the CF. As an example of the final result, we could consider the following (complete) logical chain (see Fig. 4.7).

4.3.2.2 Failure Analysis

Derivation of Failure Modes (FM) from Component Functions (CF)
Identification of Failure Causes FC (Relevant PC)

Once all the activities described above have been completed, the failure analysis can begin.

To perform optimal failure analysis, I suggest using the component functions as a starting point. First, we consider the entire functional chain of the CF (see Fig. 4.7) we want to analyze. Let's take an example by revisiting the CF discussed above.

The starting point is currently ***the failure modes***. This means we are looking for one or more errors (or failures) that can be derived from the CF in question. Let's assume the work team gives the coordinator the following technical indications (see Fig. 4.8).

Now it's time for the team to understand what the most appropriate and relevant root causes might be for the identified failure mode. Of course, we refer to

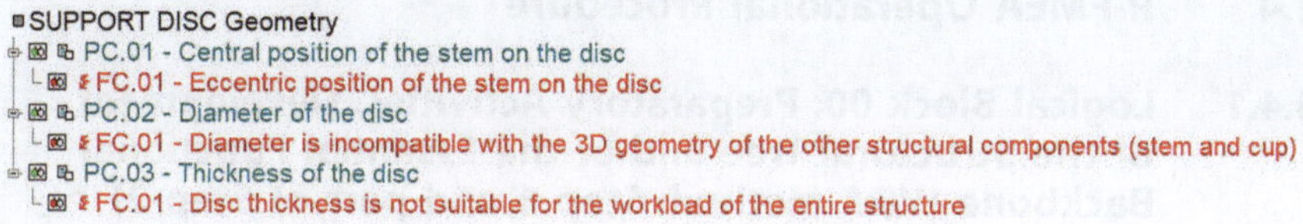

Fig. 4.9 Derivation of FCs from PCs

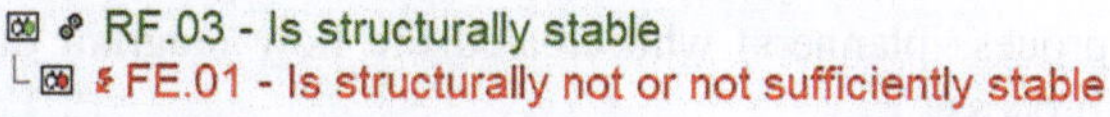

Fig. 4.10 Derivation of FEs from RFs

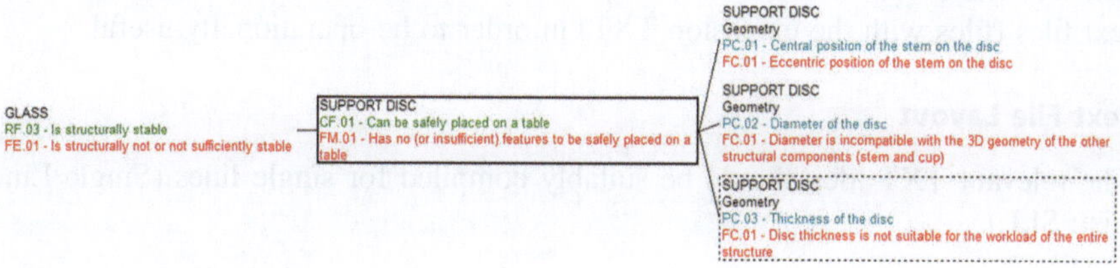

Fig. 4.11 Failure chain (complete)

the PCs we identified earlier, during the definition phase of the functional logical development.

For example, we could have (see Fig. 4.9).

A very similar investigation must be conducted regarding the failure effects. In this case, the reference functions are the product's RFs. Let's assume that we as a team have identified the following effects (see Fig. 4.10).

4.3.2.3 Failure Chains

Right Side (FC: Failure Causes)

Left Side (FE: Failure Effects)

All that remains is to ***logically link the failures***. In this case, we no longer use the function net editor, but rather the APIS-IQ failure net editor. We obtain the following final result (see Fig. 4.11).

4.4 P-FMEA Operational Procedure

4.4.1 Logical Block 00: Preparatory Activities, Development of the Structural Tree and of the Essential Functional Backbone (VDA Method: Step 1 and part of Step 2)

4.4.1.1 Input Data (Production Process Flow Chart)

The starting point for a process FMEA is the ***input data***, i.e. the data that the FMEA coordinator usually receives from his colleagues (industrial engineers or production process planners) who created the flow diagram of the industrial manufacturing process.

Data Format (Txt Text File)

The flowchart data should ideally be provided to the FMEA coordinator in simple text files (files with the extension TXT) in order to be operationally useful.

Text File Layout

The relevant TXT files should be suitably compiled for single lines (Single Line List: SLL).

4.4.1.2 Text File for Process Phase Nodes

The first text file concerns the process phase nodes, the structural elements of APIS-IQ that serve as containers for middle-level functions. In other words, this first file is nothing more than a list of the names of each process phase. Here (see Fig. 4.12) we see a simple example.

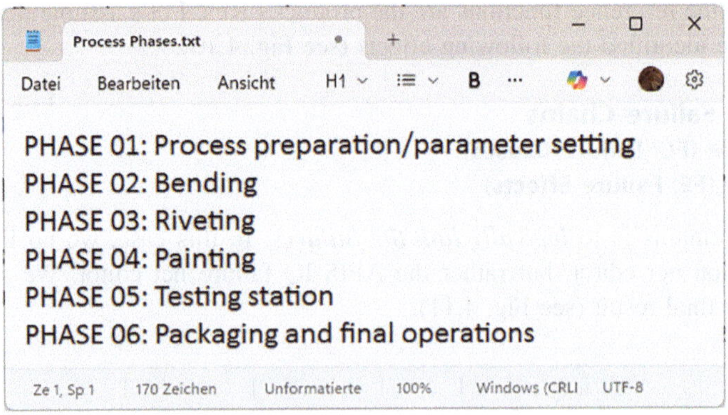

Fig. 4.12 TXT file for process phase nodes

4.4.1.3 Text File for Phase Operations (or: Phase Step Functions)

The second text file concerns ***the process phase operations***, the functional elements of APIS-IQ that describe the various activities (***PSF***:***Phase Step Functions***) planned for each phase of the production process. Let's look at a simple example in this case (see Fig. 4.13).

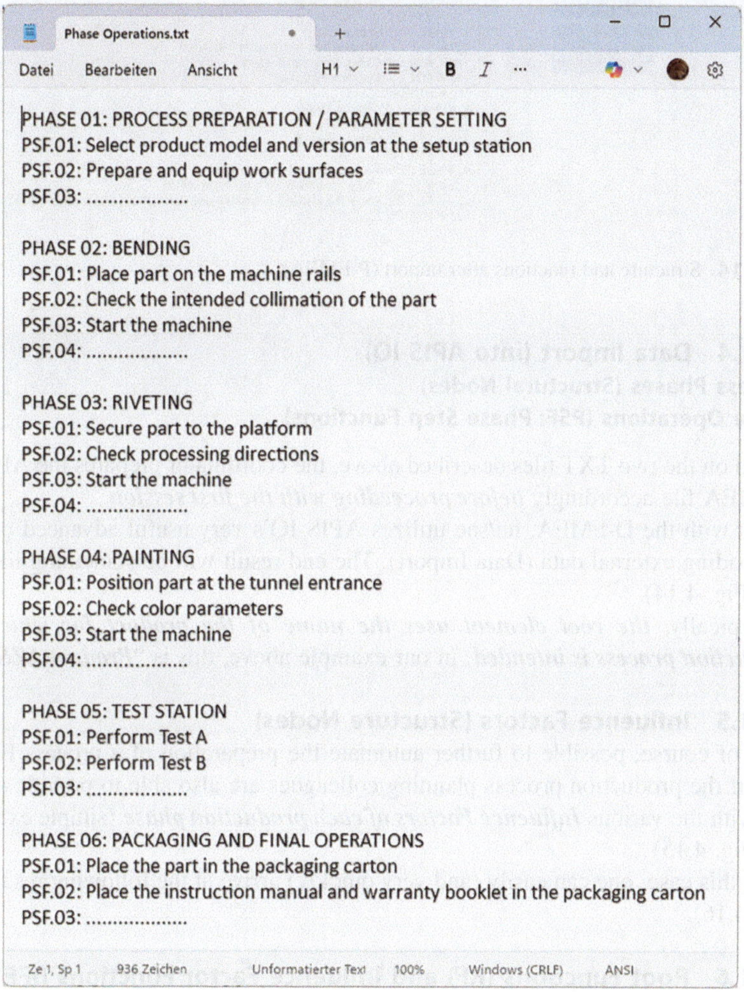

PHASE 01: PROCESS PREPARATION / PARAMETER SETTING
PSF.01: Select product model and version at the setup station
PSF.02: Prepare and equip work surfaces
PSF.03: ....

PHASE 02: BENDING
PSF.01: Place part on the machine rails
PSF.02: Check the intended collimation of the part
PSF.03: Start the machine
PSF.04: ....

PHASE 03: RIVETING
PSF.01: Secure part to the platform
PSF.02: Check processing directions
PSF.03: Start the machine
PSF.04: ....

PHASE 04: PAINTING
PSF.01: Position part at the tunnel entrance
PSF.02: Check color parameters
PSF.03: Start the machine
PSF.04: ....

PHASE 05: TEST STATION
PSF.01: Perform Test A
PSF.02: Perform Test B
PSF.03: ....

PHASE 06: PACKAGING AND FINAL OPERATIONS
PSF.01: Place the part in the packaging carton
PSF.02: Place the instruction manual and warranty booklet in the packaging carton
PSF.03: ....

Fig. 4.13 TXT file for phase operations

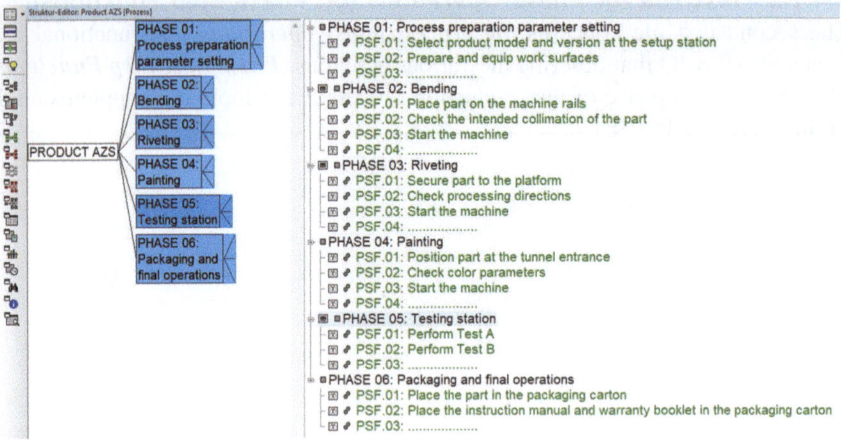

Fig. 4.14 Structure and functions after import (P-FMEA)

4.4.1.4 Data Import (into APIS-IQ)

Process Phases (Structural Nodes)

Phase Operations (PSF: Phase Step Functions)

Based on the two TXT files described above, the coordinator prepares the APIS-IQ P-FMEA file accordingly ***before proceeding with the first session.***

As with the D-FMEA, he/she utilizes APIS-IQ's very useful advanced options for loading external data (Data Import). The end result will be something like this (see Fig. 4.14).

Typically, ***the root element uses the name of the product for which the production process is intended***. In our example above, this is "***Product AZS***".

4.4.1.5 Influence Factors (Structure Nodes)

It is, of course, possible to further automate the preparation of a process FMEA, e.g., if the production process planning colleagues are also able to provide a TXT file with the various ***Influence Factors of each production phase*** (simple example, see Fig. 4.15).

In this case, one can easily (and very quickly) arrive at the following result (see Fig. 4.16).

4.4.1.6 Root Functions (RF) and Influence Factor Functions (IFF)

The ***Root Functions (RF)*** and the ***Influence Factor Functions (IFF)*** can also be usefully automated, although they are usually identified and discussed during the initial meeting of the process FMEA.

However, the relevant point here is always to ***prepare/automate the analysis as much as possible***. Otherwise, one risks not only obvious ***operational inefficiencies*** but also ***boring or giving a bad impression to colleagues from the technical departments*** by involving them in initial meetings that are actually unnecessary.

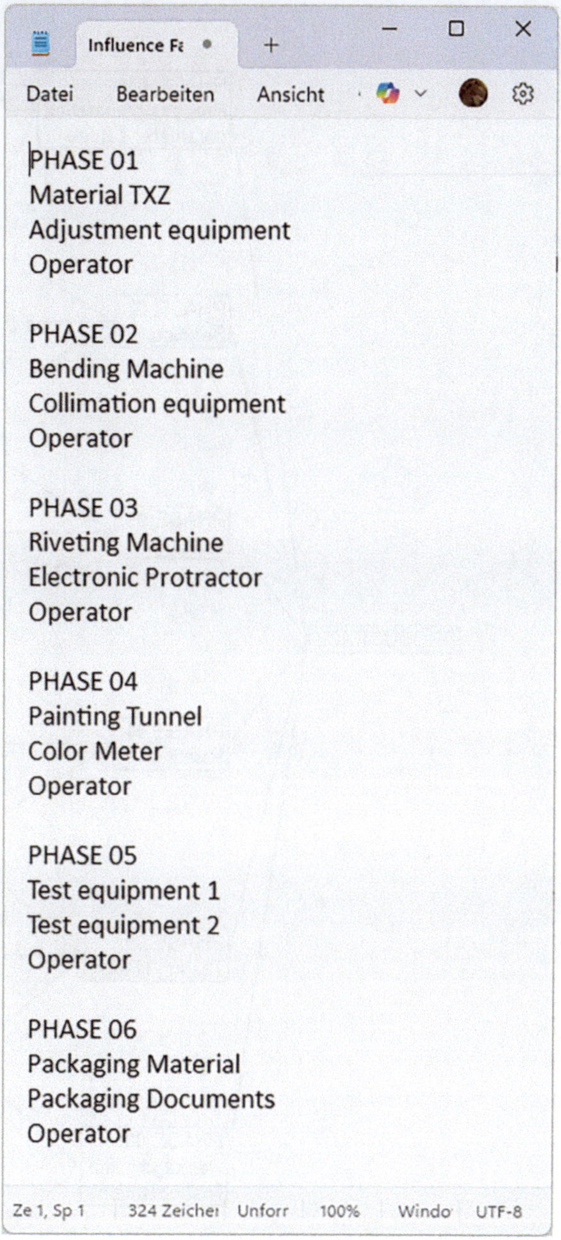

Fig. 4.15 Influence factors of a P-FMEA

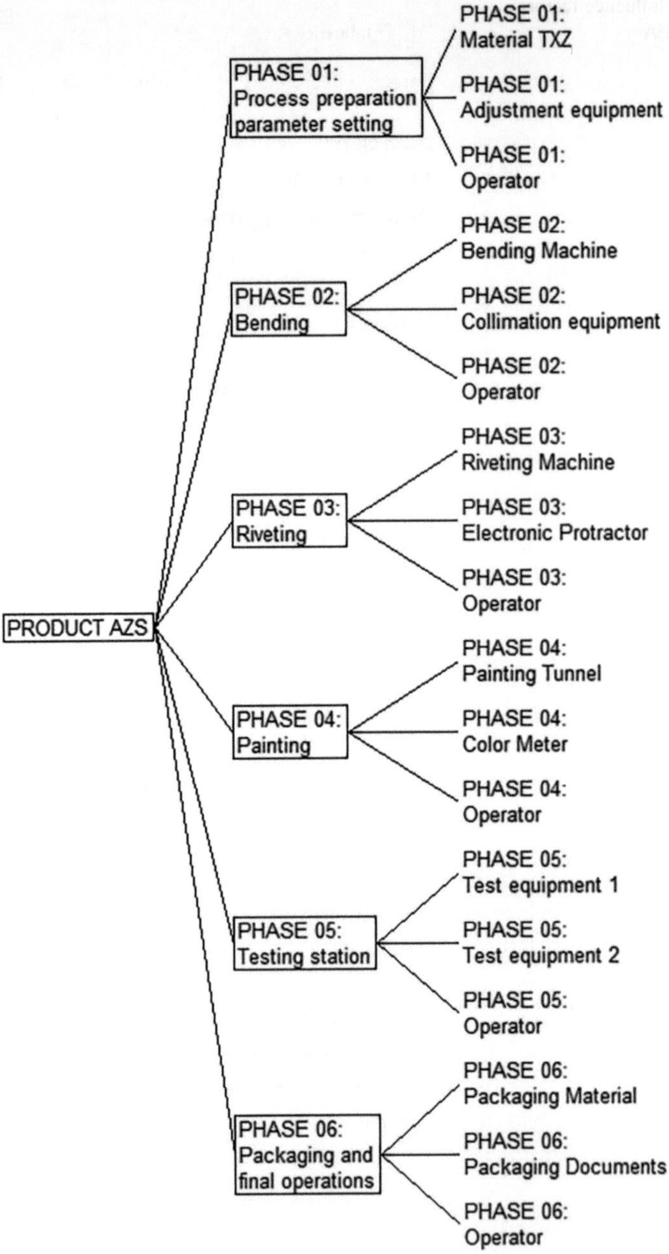

Fig. 4.16 Structure tree of a P-FMEA (complete)

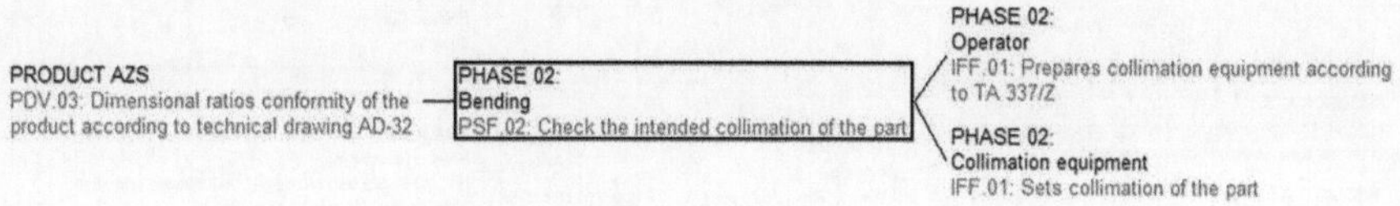

Fig. 4.17 Logical connections (function chain)

My suggestion in this context is to encourage colleagues from the technical departments to identify (if possible) the ***entire functional framework*** before starting the teamwork, including a set (at least minimal) of ***Product and Process Variables*** and ***Influence Factor Functions***, which are then assigned to the appropriate structural nodes (RE and IF, respectively).

The development of logical connections, i.e. the logical cause-effect relationships that connect the various functions (see Fig. 4.17).

can then form the ***actual (and fruitful) starting point for teamwork (first FMEA meeting)***. In the figure we see a simple example of such a function network.

4.4.2 Logical Block 01: System Analysis (VDA Method: Step 2 and Complete Step 3)

4.4.2.1 Functional Chains

Right Side (Influence Factors/influence Factor Functions)

Left Side (Root Element: Product and Process Variables)

After all the activities planned in logical block 00 have been performed, we obtain ***a complete structure tree up to the phase operations.***

We now need to complete the functional analysis provided in the second step of the VDA method. This means that for each specific phase operation, we need to:

- Identify all relevant Influence Factors (IF) and the corresponding Influence Factor Functions (IFF)
- Link the influence factor functions to the phase operations (right side of the function chain)
- Identify all relevant root functions (i.e., product and/or process variables) listed under the root element
- Link the root functions to the operation (left side of the function chain)

Let's give an example here. Suppose we need to handle the following operation:

- ***PHASE 03/PSF.04: Rivet the base plate 12 times according to specifications***

First, we should ask ourselves ***some technically specific questions***:

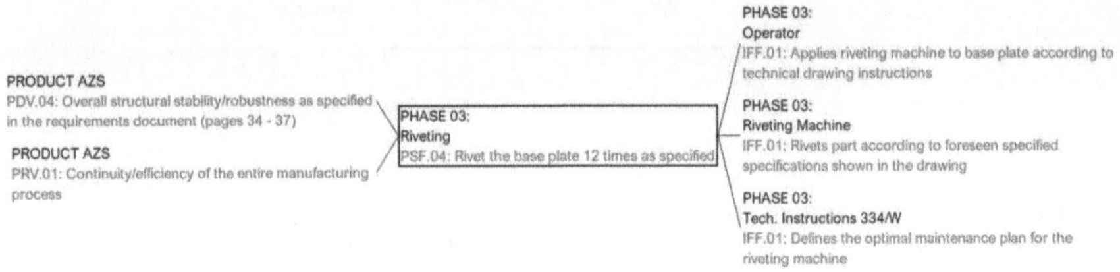

Fig. 4.18 Function chain of a P-FMEA (complete)

- How many and which influence factors should be correctly considered here?
- Even better: ***Which influence factors and through which specialized factor functions could best explain and justify the phase operation we are dealing with (i.e., according to the strictest possible cause-and-effect logic)?***

In this case, an appropriate technical response could be:

- ***PHASE 03/OPERATOR/IFF.01***: *Applies riveting machine to base plate according to technical drawing instructions*
- ***PHASE 03/RIVETING MACHINE/IFF.01***: *Rivets part according to specified specifications as per drawing*
- ***PHASE 03/TECHNICAL INSTRUCTION 334W/IFF.01***: *Defines optimal maintenance plan for the riveting machine*

We therefore have ***3 influence factors*** with ***3 corresponding specialized functions of these factors***.

- ***MAN*** (Operator)
- ***MACHINE*** (Riveting machine)
- ***METHOD*** (Technical instructions)

Now all that remains is the left side of the chain, i.e., the second (and final) logical cause-and-effect relationship. To achieve this result, we proceed exactly as before. We therefore first identify the relevant root functions under the root element and link them together with the operation.

As an example of the final result, we could consider the following logical concatenation (complete) (see Fig. 4.18).

The abbreviations **"*PDV*" *and* "*PRV*"**, which appear on the left side of the chain as specific identifiers of the relevant root functions, mean:

- ***PDV***: ***Product variable***
- ***PRV***: ***Process variable***

exactly as we already defined at the beginning of Sect. 4.4.

4.4.2.2 Preliminary Analyses: Specification and Failure Relevance Analysis (Through Creation of APIS-IQ-Uda: User-Defined Attributes)

At this point in the procedure, we should conduct ***two essential*** "***preliminary analyses***":

- A ***first preliminary analysis (TS preliminary analysis)*** to identify all phase operations for which ***a technical specification is foreseen***
- A ***second preliminary analysis—FR preliminary analysis (FR***: ***Failure Relevance)***—to understand which operations would essentially not require an in-depth FMEA because, for example, they are not technically complex and/ or have never historically been identified as problematic

These two preliminary analyses would have the purpose of facilitating and reducing the burden on the entire P-FMEA by making it leaner and faster, since (possibly several) operations, with their corresponding logical chains.

- could be treated in a standardized manner in the ***TS pre-analysis*** with regard to the failures, and
- could be excluded based on the ***FR pre-analysis*** from the FMEA (according to the technical judgment of the team, of course)

To achieve this, it would be advisable to use the so-called "***User-defined Attributes***", which in this context would undoubtedly provide immediate and very useful visual support in APIS-IQ.

These attributes are essentially ***markers*** that can be advantageously applied to many objects, especially functions and failures. For example, the APIS-IQ window (see Fig. 4.19) for the properties of the operation "***PSF.04***: ***Rivet the base plate according to specification 12 times***", which we have already seen above, would look like this after defining the user attributes "TS" and "FR":

By checking the relevant boxes, visual results such as the following (see Fig. 4.20) could then be obtained:

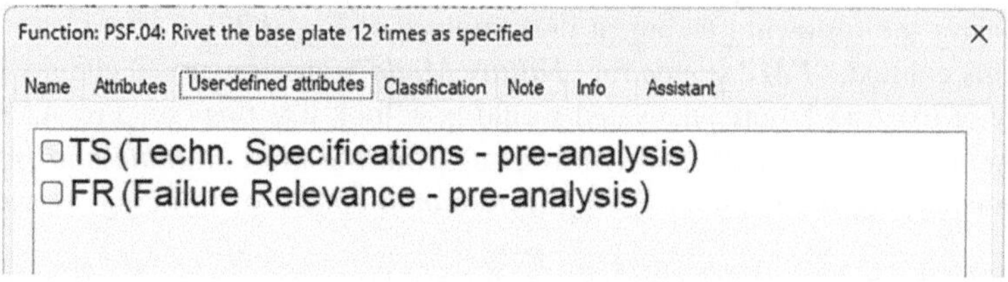

Fig. 4.19 APIS-IQ Window "User-defined attributes TS and FR"

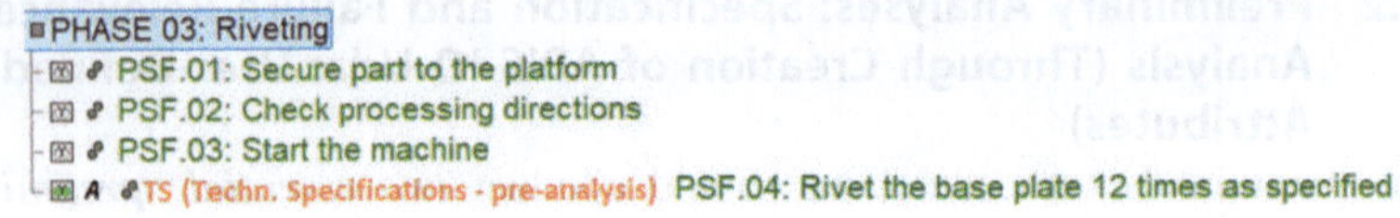

Fig. 4.20 View after selecting UD attributes

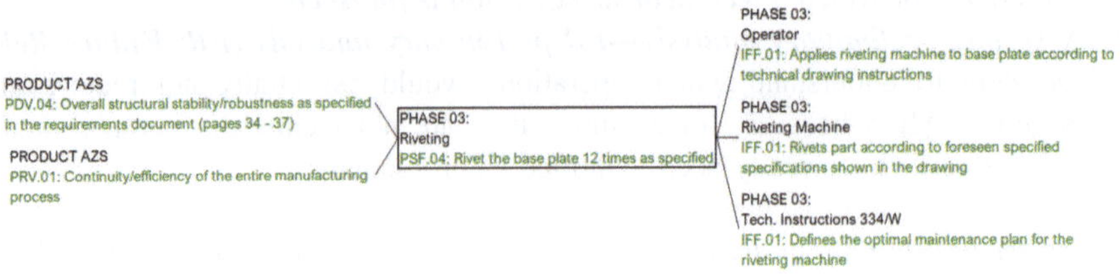

Fig. 4.21 Functional chain of the operation to be analyzed

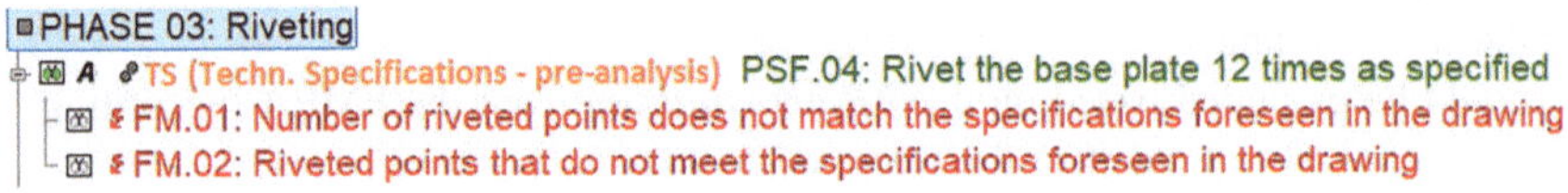

Fig. 4.22 Failure modes derived from a phase operation

4.4.2.3 Failure Analysis

Derivation of Failure Modes (FM) from Phase Operations (PSF)
Identification of Failure Causes (Relevant Influence Factors)

After all the activities described above have been carried out, the ***Failure Analysis*** can be started.

To perform an optimal failure analysis, I would suggest ***taking the operations as a starting point***. First, we consider the entire functional chain (see Fig. 4.21) of the operation we want to analyze. Let's take the operation discussed above as an example:

The starting point is currently the failure modes. This means that we are looking for one or more errors (or failures) that can be derived from the function in question (phase operation or phase step function). Let's assume that the team gives the coordinator the following technical indications (see Fig. 4.22).

In this context, "***FM***" stands for "***Failure Mode***". The functional chain with the display of existing failures activated would now look like this (see Fig. 4.23).

Now it's time to understand (as a team) ***what the most suitable and relevant failure causes of the two identified failure modes could be***. In doing so, we will,

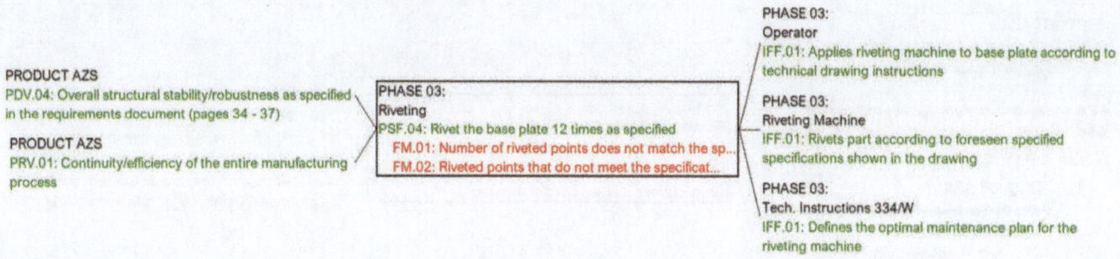

Fig. 4.23 Function chain with activated display of failure modes

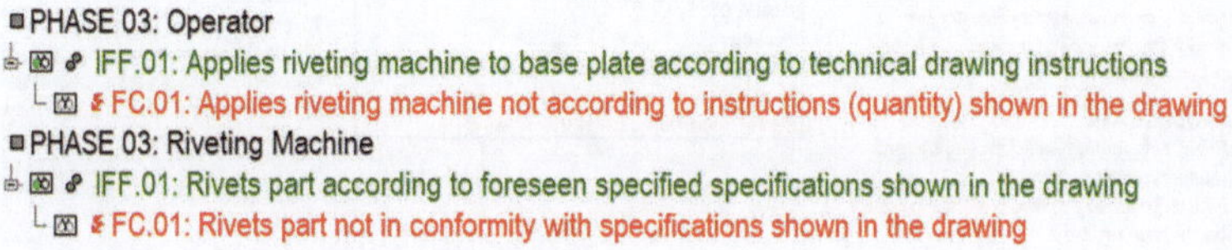

Fig. 4.24 Identifying the failure causes in IFs/IFFs

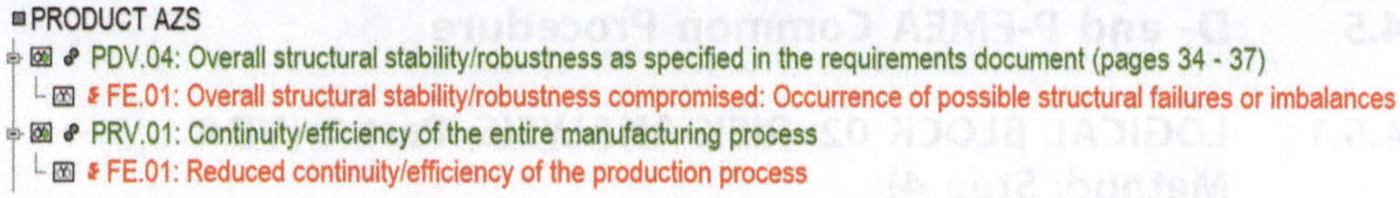

Fig. 4.25 Identification of failure effects in RE/PDVs and RE/PRVs

of course, refer to the already identified specialized functions that were defined in the function framework definition phase.

Here, for example, we could have (see Fig. 4.24).

A similar analysis must be carried out for the ***failure effects***. In this case, the reference functions are the ***product/process variables***. Let's assume that the team has identified the following effects (see Fig. 4.25).

4.4.2.4 Failure Chains
Right Side (Failure Causes)
Left Side (Failure Effects)

All that now remains is to ***logically link the failures together***. In this case, we no longer use the function network editor, but the failure network editor from APIS-IQ. We obtain the following two final results (see Fig. 4.26).

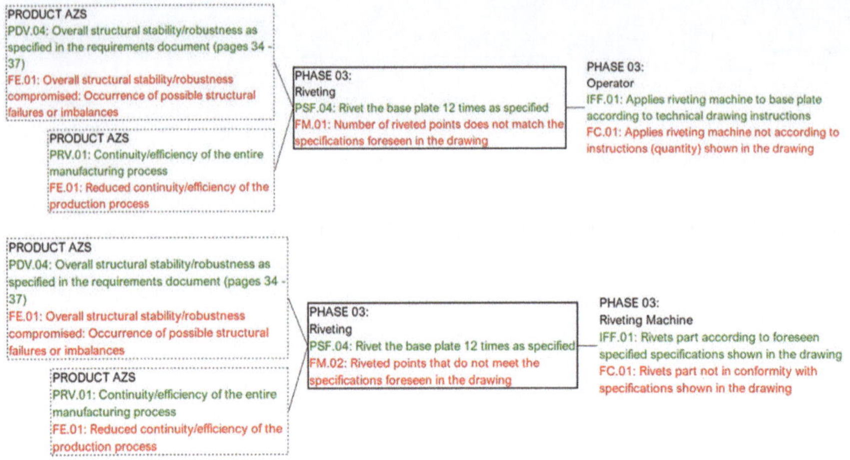

Fig. 4.26 Example of failure chains (complete)

4.5 D- and P-FMEA Common Procedure

4.5.1 LOGICAL BLOCK 02: RISK ANALYSIS, Part 1 (VDA Method: Step 4)

4.5.1.1 Preparation of the Severity Evaluation

Following the failure analysis, the VDA method foresees the so-called "form analysis", which involves conducting the action analysis and the risk assessment (evaluation of the ***three risk parameters: severity, occurrence, and detection***).

Various activities related to the form analysis can (and should, for operational efficiency) be performed before opening the forms (activities that precede the form analysis).

The first of these is the ***severity assessment***, i.e., ***assigning values (from 1 to 10, as for the other two parameters) to all failure consequences*** related to the various root functions. It is advisable to perform this entire activity directly in the APIS-IQ structure list (the field to the right of the field where the structure tree is displayed and managed). Minimal examples are shown below—see Fig. 4.27 (D-FMEA) and Fig. 4.28 (P-FMEA).

The corresponding ***VDA evaluation tables***—see Fig. 4.29 (D-FMEA) and Fig. 4.30 (P-FMEA)—look like this:

4.5.1.2 Formalization of the Work Team

A second preparatory activity for form analysis is the ***formulation (system entry) of the work team***. In this context, we can usefully use the APIS-IQ editor, which is specifically designed for this function.

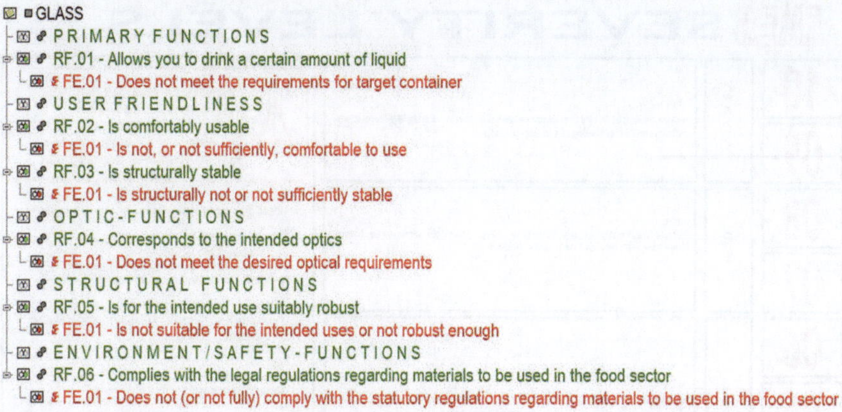

Fig. 4.27 Severity evaluation D-FMEA

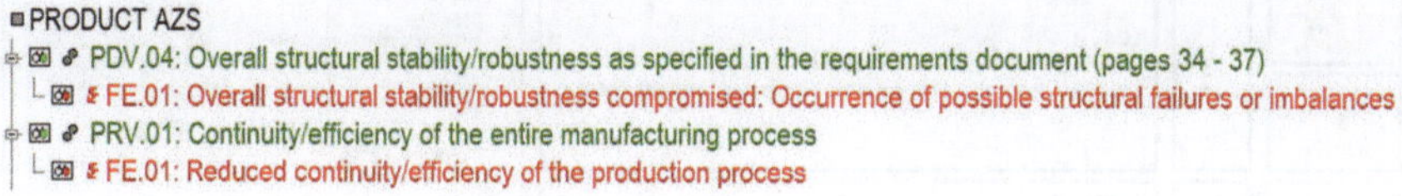

Fig. 4.28 Severity evaluation P-FMEA

Let's look at an example (see Fig. 4.31) of what the formulation of a work team would look like (the image refers to a D-FMEA, but would be very similar for a P-FMEA).

Please note that the members of the (entire) work team were actually assigned to ***two different teams***:

- ***CORE Team***: groups the members whose participation is essentially required for each session of the analysis cycle
- ***ENLARGED Team***: groups the members whose participation depends on the specific technical topics that are addressed during the various sessions of the analysis cycle

4.5.1.3 Definition of Symbolic Deadlines (Action Statuses and Milestones)

A third preparatory activity for the form analysis is the ***definition of "symbolic deadlines"*** related to two APIS-IQ objects:

- ***Action states.***
- ***Action deadlines*** (milestones).

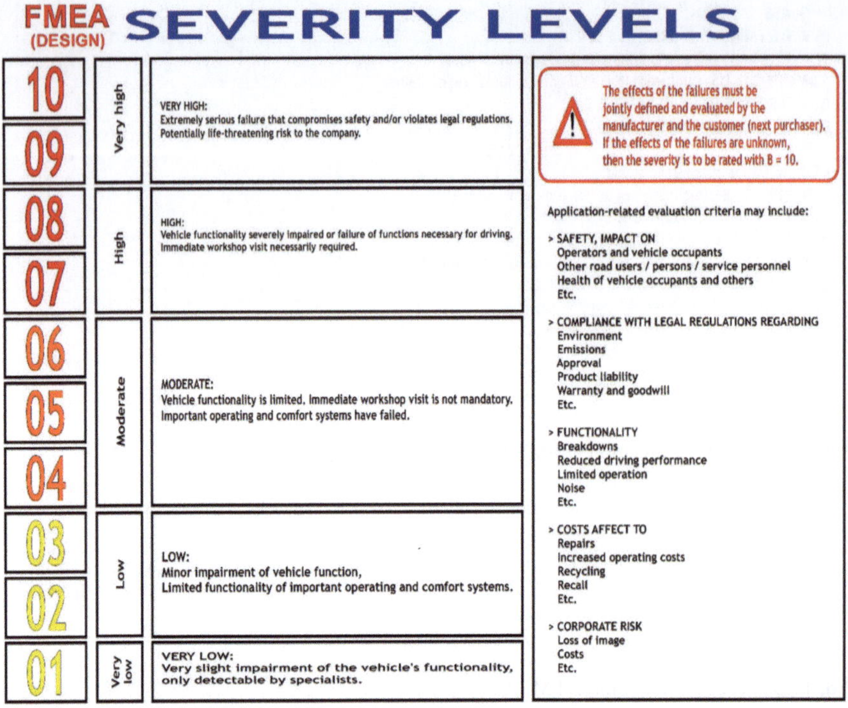

Fig. 4.29 VDA Evaluation table SEVERITY (D-FMEA)

The "***milestones***" are deadlines that are usually already defined in the project schedule and which the design or production process refers to. Symbolic deadlines are also managed in this case using a dedicated editor in APIS-IQ. Let's look at an example (see Fig. 4.32—the image refers to a D-FMEA, but would be very similar for a P-FMEA).

As the example illustrates, specifying a concrete calendar date, which is then assigned to the various symbolic dates, is possible but not mandatory.

As with all APIS-IQ editors, ***the major advantage of this centralized approach is that any changes to the parameters managed by the editor can ONLY be processed effectively within the editor itself***. ***All forms***, as well as all possible external contexts related to the editor, are ***automatically updated***. This results in an undeniable and generally highly valued increase in operational efficiency.

4.5.1.4 Definition of the Risk Matrix (Optimization criterion—JI Matrix)

Another very important point before opening the forms is ***defining an optimization criterion***, which is then used in the fifth step of the method, i.e., at the end of the form analysis.

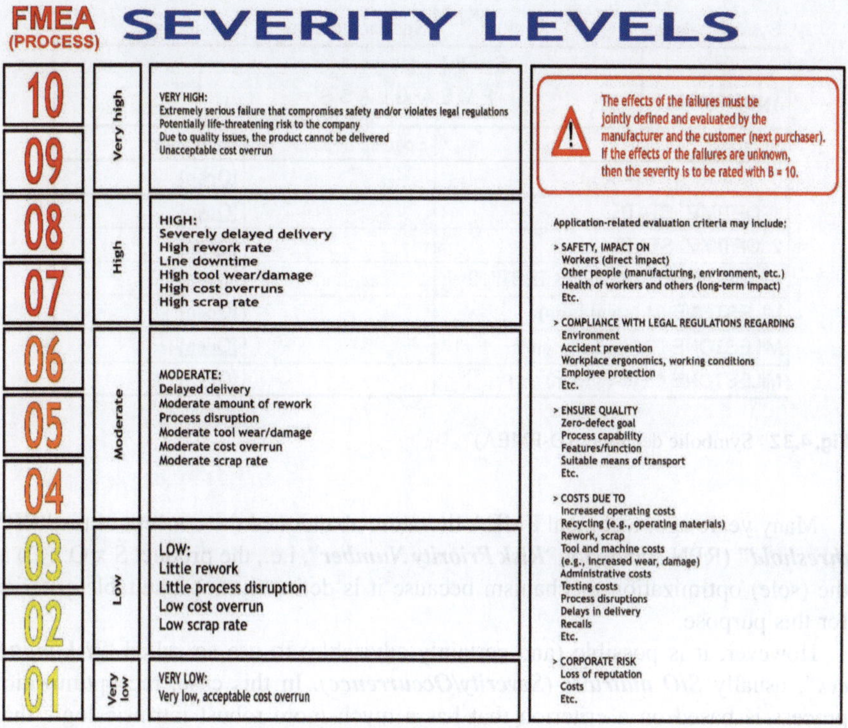

Fig. 4.30 VDA Evaluation table SEVERITY (P-FMEA)

Team	Name /	First name	Department
INDUSTRIAL	Teams and persons D-FMEA GLASS		
Team	Name /	First name	Department
Team: ENLARGED TEAM (2 Persons)			
ENLARGED TEAM	Mahler	Peter	Pre-development
	Supervisor		P. Luigi Belcaro FMEA Coordinator
Team: CORE TEAM (4 Persons)			
CORE TEAM	Klein	Hannes	Design
	Müller	Wilhelm	Design
	Reddig	Carsten	Industrial Engineering
	Supervisor		P. Luigi Belcaro FMEA Coordinator

Fig. 4.31 Example work team (D-FMEA)

Symbolic deadline	Specific deadline	Instances
INDUSTRIAL	Symbolic deadlines D-FMEA GLASS	
Symbolic deadline	Specific deadline	Instances
0 - CURRENT STATE		[Open]
1. OPTIMIZ. STATE		[Open]
2. OPTIMIZ. STATE		[Open]
AFTER EFFECTIVENESS TESTING		[Open]
MILESTONE 01 (short term)		[Open]
MILESTONE 02 (medium term)		[Open]
MILESTONE 03 (long term)		[Open]

Fig. 4.32 Symbolic deadlines (D-FMEA)

Many years ago, technical FMEA literature abandoned the concept of the "***RPN threshold***" (RPN stands for "***Risk Priority Number***", i.e., the product S x O x D) as the (sole) optimization mechanism because it is definitely not a suitable criterion for this purpose.

However, it is possible (and certainly advisable) to use so-called "***risk matrices***", usually ***S/O matrices (Severity/Occurrence).*** In this case, the optimization process is based on a criterion that has a much more robust intrinsic logic than that associated with an RPN threshold.

Of all the possible S/O matrices currently available in the best technical FMEA literature, I personally always recommend using the "Lindland matrix," as its development is based on particularly robust and comprehensible principles (see the book: "The 7 Failure Modes," by John L. Lindland, The Bella Group Editions, page 176).

The matrix in question (see Fig. 4.33) looks like this:

4.5.1.5 Creation of the Forms

Before starting the form analysis, ensure that the forms are opened ***in the correct number and at the appropriate structural points*** (i.e., at the "correct nodes" of the structure tree).

In this context, it suffices to note that—since a form is an APIS-IQ "FM-centric" object, i.e., centered on failure modes—***the only nodes from which the forms should be created are the component (D-FMEA) or phase (P-FMEA) nodes***. In other words:

a ***D-FMEA***	whose product for example, of	***n components***	Consists
a ***P-FMEA***		***n process phases***	

Will also Have N Forms.

CRITICALITY: S x O

OCCURRENCE

O/S	1	2	3	4	5	6	7	8	9	10
10	10	20	30	40	50	60	70	80	90	100
9	9	18	27	36	45	54	63	72	81	90
8	8	16	24	32	40	48	56	64	72	80
7	7	14	21	28	35	42	49	56	63	70
6	6	12	18	24	30	36	42	48	54	60
5	5	10	15	20	25	30	35	40	45	50
4	4	8	12	16	20	24	28	32	36	40
3	3	6	9	12	15	18	21	24	27	30
2	2	4	6	8	10	12	14	16	18	20
1	1	2	3	4	5	6	7	8	9	10

SEVERITY

John Lindland's matrix

Fig. 4.33 Optimization matrix by John Lindland

4.5.1.6 Working with Forms

Creation of a Column for the Risk Matrix (Jl-M Column)

When working with a form, it is always very useful to prepare ***a column beforehand for the immediate display of the status of each row (corresponding to a single failure cause) in the risk matrix (optimization criterion)***.

In this way, you can continuously monitor the status of each failure cause row by row (or analogously, failure cause by failure cause) with regard to the future (next step of the VDA method) potential need for optimization.

Let's look at ***an example of what a form looks like***—see Fig. 4.34 (D-FMEA) and Fig. 4.35 (P-FMEA).

The ***sixth column (from the left) above (column JL-M)*** is precisely the column that allows us to display the optimization status of a specific failure cause in advance.

For now, we will limit ourselves to noting just one important aspect: the ***layout of the form***. In this case, we have (deliberately) chosen a ***VDA layout***, meaning ***a layout that corresponds to the typical positioning of FMEA objects in a structure fully compliant with the VDA standard***.

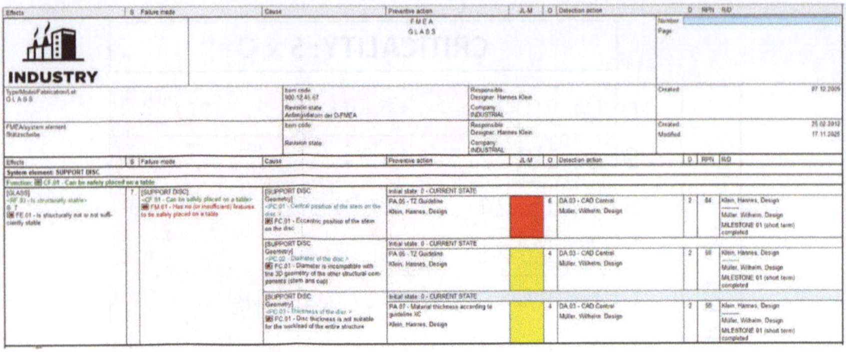

Fig. 4.34 JL-M column (D-FMEA form)

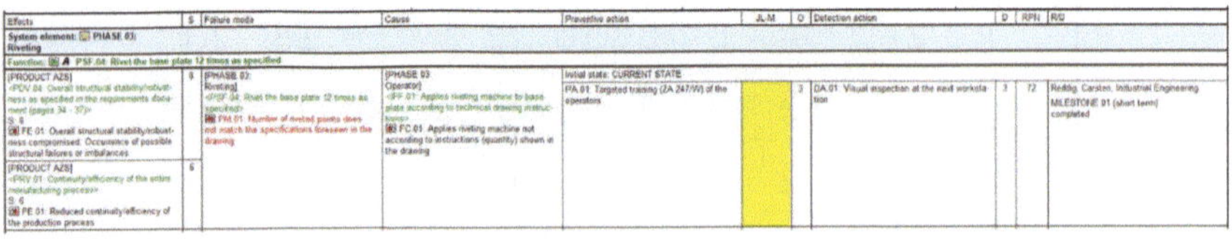

Fig. 4.35 JL-M column (P-FMEA form)

Clear and Unequivocal Definition of <current> in the Action Statuses

It is particularly important that ***the work team is clearly aware of what "current" really means in the context of the ongoing FMEA.***

More precisely:

Current status (of the actions) means that ***the described actions document everything that is already present (has actually been/is being implemented) in product development or the manufacturing process at the time the analysis is performed.***

We will see a number of concrete examples shortly.

Definition of Preventive Actions

To define a suitable ***Preventive Action*** (whose corresponding APIS-IQ object we identify with the marker "PA" followed by a period and a sequential number), it is helpful to ask the work team the following technically targeted question:

Given a specific failure cause, what could/would we do with the best and most modern development and/or production process technologies available to us to PREVENT the occurrence of the failure in question?

In other words:

What technically targeted actions would allow us to reasonably claim that their implementation would actually PREVENT the occurrence of the problems associated with the failure cause in question?

Effects	S	Failure mode	Cause	Preventive action
System element: SUPPORT DISC				
Function: CF.01 - Can be safely placed on a table				
[GLASS] <RF.03 - Is structurally stable> S: 7 FE.01 - Is structurally not or not sufficiently stable	7	[SUPPORT DISC] <CF.01 - Can be safely placed on a table> FM.01 - Has no (or insufficient) features to be safely placed on a table	[SUPPORT DISC Geometry] <PC.01 - Central position of the stem on the disc > FC.01 - Eccentric position of the stem on the disc	Initial state: 0 - CURRENT STATE PA.05 - TZ Guideline Klein, Hannes, Design

Fig. 4.36 Example of preventive action (PA)—D-FMEA

Effects	S	Failure mode	Cause	Preventive action
System element: PHASE 03: Riveting				
Function: *A* PSF.04: Rivet the base plate 12 times as specified				
[PRODUCT AZS] <PDV.04: Overall structural stability/robustness as specified in the requirements document (pages 34 - 37)> S: 8 FE.01: Overall structural stability/robustness compromised: Occurrence of possible structural failures or imbalances	8	[PHASE 03: Riveting] <PSF.04: Rivet the base plate 12 times as specified> FM.01: Number of riveted points does not match the specifications foreseen in the drawing	[PHASE 03: Operator] <IFF.01: Applies riveting machine to base plate according to technical drawing instructions> FC.01: Applies riveting machine not according to instructions (quantity) shown in the drawing	Initial state: CURRENT STATE PA.01: Targeted training (ZA 247/W) of the operators
[PRODUCT AZS] <PRV.01: Continuity/efficiency of the entire manufacturing process> S: 6 FE.01: Reduced continuity/efficiency of the production process	6			

Fig. 4.37 Example of preventive action (PA)—P-FMEA

Let's look at an example—see Fig. 4.36 (D-FMEA) and Fig. 4.37 (P-FMEA). In this case, we are considering a ***preventive measure***:

- a ***design guideline (D-FMEA)***
- a ***targeted employee training (P-FMEA)***

which explicitly aims to (reasonably) ensure that

- ***the base of the cup stem is located precisely in the center of the support disc***
- ***the operator's technical operations with the riveting machine are carried out exactly as explicitly specified in the instructions (TA 297/EL)***

Occurrence Evaluation

The next step is for the team to define an ***occurrence level*** for the investigated failure cause (accompanied by the preventive action just defined). Here is the precise technical question that should be posed to the team:

With which probability, considering and applying the preventive action just defined (note: conditional, not free probability), can we assume that the failure cause in question would occur?

Here—based on the risk parameter rating tables (all on a scale from 1 to 10, as already mentioned) provided by each standard—the team will give its assessment to the coordinator. For example, see Fig. 4.38 (D-FMEA) and Fig. 4.39 (P-FMEA).

Effects	S	Failure mode	Cause	Preventive action	JL-M	O
System element: SUPPORT DISC						
Function: CF.01 - Can be safely placed on a table						
[GLASS] <RF.03 - Is structurally stable> S: 7 FE.01 - Is structurally not or not sufficiently stable	7	[SUPPORT DISC] <CF.01 - Can be safely placed on a table> FM.01 - Has no (or insufficient) features to be safely placed on a table	[SUPPORT DISC Geometry] <PC.01 - Central position of the stem on the disc > FC.01 - Eccentric position of the stem on the disc	Initial state: 0 - CURRENT STATE PA.05 - TZ Guideline Klein, Hannes, Design		6

Fig. 4.38 Example of occurrence evaluation—D-FMEA

Effects	S	Failure mode	Cause	Preventive action	JL-M	O
System element: PHASE 03: Riveting						
Function: A PSF.04: Rivet the base plate 12 times as specified						
[PRODUCT AZS] <PDV.04: Overall structural stability/robustness as specified in the requirements document (pages 34 - 37)> S: 8 FE.01: Overall structural stability/robustness compromised: Occurrence of possible structural failures or imbalances	8	[PHASE 03: Riveting] <PSF.04: Rivet the base plate 12 times as specified> FM.01: Number of riveted points does not match the specifications foreseen in the drawing	[PHASE 03: Operator] <IFF.01: Applies riveting machine to base plate according to technical drawing instructions> FC.01: Applies riveting machine not according to instructions (quantity) shown in the drawing	Initial state: CURRENT STATE PA.01: Targeted training (ZA 247/W) of the operators		3
[PRODUCT AZS] <PRV.01: Continuity/efficiency of the entire manufacturing process> S: 6 FE.01: Reduced continuity/efficiency of the production process	6					

Fig. 4.39 Example of occurrence evaluation—P-FMEA

The ***VDA product and process evaluation tables*** provide the following for the parameter ***occurrence*** to support the identification of the applicable occurrence levels—see Fig. 4.40 (D-FMEA) and Fig. 4.41 (P-FMEA).

Please note that in our forms, the JL-M column has got the ***color red (D-FMEA) or yellow (P-FMEA)***, corresponding to the failure causes currently being evaluated. This is because the Lindland matrix (selected optimization criterion) assigns to the following combinations:

- (B, A) = (7, 6) (D-FMEA)
- (B, A) = (8, 3) (P-FMEA)

the same colors.

As mentioned earlier, this visual aid is particularly useful for the team to immediately gain an overview of the ***optimization possibilities for the failure causes being analyzed*** (individual rows of a form).

Definition of Detection Actions

To properly define a ***Detection Action*** (whose corresponding APIS-IQ object we will identify with the marker "DA," followed by a period and a sequential number), it is helpful to pose the following technically targeted question to the work team:

Suppose there is a specific failure cause, what could/would we do, using the best and most advanced development and/or production process technologies available to us, to DETECT the occurrence of the failure cause every time it manifests?

In other words:

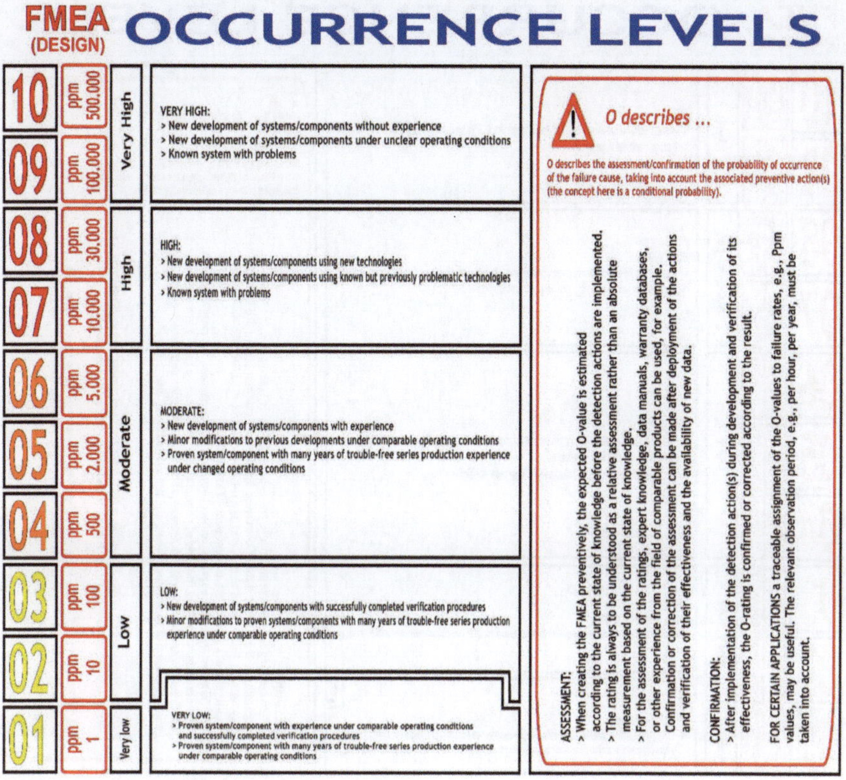

Fig. 4.40 VDA Occurrence table—D-FMEA

What technically targeted actions (control mechanisms, in this case) would allow us to reasonably claim that we would detect the occurrence of the problems associated with the failure cause in every instance BEFORE anyone downstream of the development and/or production process (typically the end customer) notices it? Let's look at a couple of examples—see Figs. 4.42 (D-FMEA) and 4.43 (P-FMEA).

In this case, the detection action ***(D-FMEA)*** corresponds to a ***CAD control*** that the work team considers sufficiently robust to prevent the problem in question from escaping the designer's attention.

Regarding the P-FMEA example, from the perspective of overall production process efficiency, one could discuss placing the relevant ***control at a workstation downstream*** of the station to which the operation and the corresponding failure cause relate.

When deciding on actions, factors such as their cost and their cost–benefit ratio (CBR) should also be considered. The CBR also reflects the historically significant impact of certain technical problems.

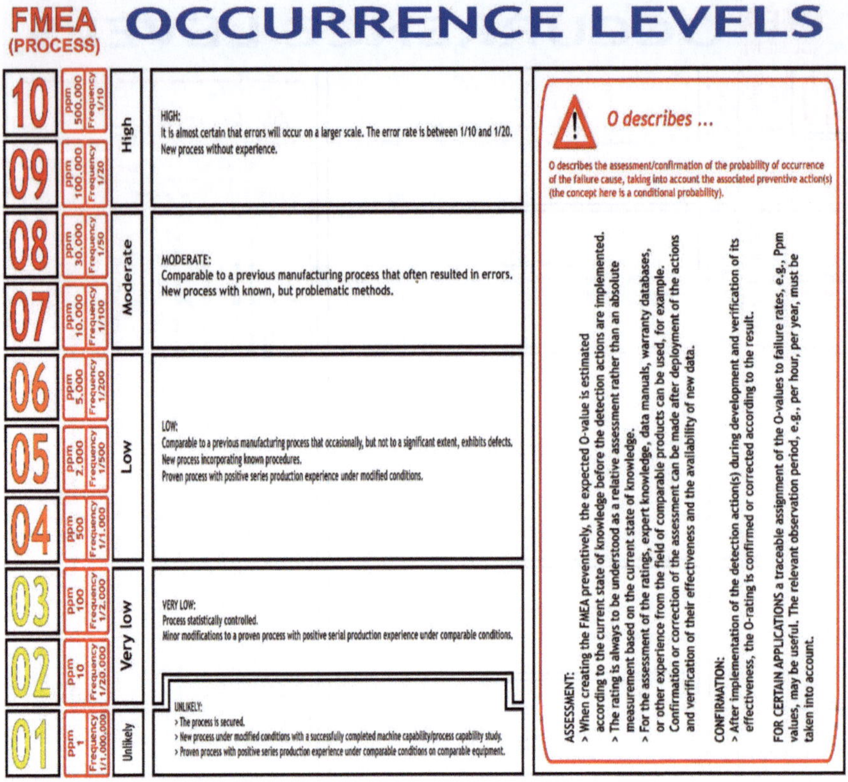

Fig. 4.41 VDA Occurrence table—P-FMEA

Effects	S	Failure mode	Cause	Preventive action	JL-M	O	Detection action
System element: SUPPORT DISC							
Function: CF.01 - Can be safely placed on a table							
				Initial state: 0 - CURRENT STATE			
[GLASS] <RF.03 - Is structurally stable> S: 7 FE.01 - Is structurally not or not sufficiently stable	7	[SUPPORT DISC] <CF.01 - Can be safely placed on a table> FM.01 - Has no (or insufficient) features to be safely placed on a table	[SUPPORT DISC Geometry] <PC 01 - Central position of the stem on the disc > FC.01 - Eccentric position of the stem on the disc	PA.05 - TZ Guideline Klein, Hannes, Design		6	DA.03 - CAD Control Müller, Wilhelm, Design

Fig. 4.42 Example of detection action (DA)—D-FMEA

Effects	S	Failure mode	Cause	Preventive action	JL-M	O	Detection action
System element: PHASE 03: Riveting							
Function: *A* PSF.04: Rivet the base plate 12 times as specified							
				Initial state: CURRENT STATE			
[PRODUCT AZS] <PDV.04: Overall structural stability/robustness as specified in the requirements document (pages 34 - 37)> S: 8 FE.01: Overall structural stability/robustness compromised: Occurrence of possible structural failures or imbalances	8	[PHASE 03: Riveting] <PSF.04: Rivet the base plate 12 times as specified> FM.01: Number of riveted points does not match the specifications foreseen in the drawing	[PHASE 03: Operator] <IFF.01: Applies riveting machine to base plate according to technical drawing instructions> FC.01: Applies riveting machine not according to instructions (quantity) shown in the drawing	PA.01: Targeted training (ZA 247/W) of the operators		3	DA.01: Visual inspection at the next workstation
[PRODUCT AZS] <PRV.01: Continuity/efficiency of the entire manufacturing process> S: 6 FE.01: Reduced continuity/efficiency of the production process	6						

Fig. 4.43 Example of detection action (DA)—P-FMEA

Effects	S	Failure mode	Cause	Preventive action	JL-M	O	Detection action	D	RPN
System element: SUPPORT DISC									
Function: CF.01 - Can be safely placed on a table									
[GLASS] <RF.03 - Is structurally stable> S: 7 FE.01 - Is structurally not or not sufficiently stable	7	[SUPPORT DISC] <CF.01 - Can be safely placed on a table> FM.01 - Has no (or insufficient) features to be safely placed on a table	[SUPPORT DISC Geometry] <PC.01 - Central position of the stem on the disc > FC.01 - Eccentric position of the stem on the disc	Initial state: 0 - CURRENT STATE PA.05 - TZ Guideline Klein, Hannes, Design		6	DA.03 - CAD Control Müller, Wilhelm, Design	2	84

Fig. 4.44 Example of detection evaluation—D-FMEA

Effects	S	Failure mode	Cause	Preventive action	JL-M	O	Detection action	D	RPN
System element: PHASE 03: Riveting									
Function: A PSF.04: Rivet the base plate 12 times as specified									
[PRODUCT AZS] <PDV.04: Overall structural stability/robustness as specified in the requirements document (pages 34 - 37)> S: 8 FE.01: Overall structural stability/robustness compromised: Occurrence of possible structural failures or imbalances	8	[PHASE 03: Riveting] <PSF.04: Rivet the base plate 12 times as specified> FM.01: Number of riveted points does not match the specifications foreseen in the drawing	[PHASE 03: Operator] <IFF.01: Applies riveting machine to base plate according to technical drawing instructions> FC.01: Applies riveting machine not according to instructions (quantity) shown in the drawing	Initial state: CURRENT STATE PA.01: Targeted training (ZA 247/W) of the operators		3	DA.01: Visual inspection at the next workstation	3	72
[PRODUCT AZS] <PRV.01: Continuity/efficiency of the entire manufacturing process> S: 6 FE.01: Reduced continuity/efficiency of the production process	6								

Fig. 4.45 Example of detection evaluation—P-FMEA

Detection Evaluation

The next step is for the team to define a ***detection level*** for the investigated failure cause (accompanied by the detection action just defined). Here is the precise technical question that should be posed to the team:

Given the assumption and application of the detection action just defined (conditional probability), what is the probability that the failure cause in question would NOT be detected?

Or similarly:

How "LOW in Power" (in Terms of Reliable and Effective Detection, of Course) is the Detection Action in Question?

Note that I have deliberately emphasized here:

- ***The failure cause in question would NOT be detected***
- ***How "LOW in power"***

Because ***in this case, the assessment is "inverted"*** compared to the usual way of using the scale 1–10. This happens for the sake of maintaining the usual "reading direction":

- ***01**: **very low risk (negligible, almost imperceptible risk)***
- ***10**: **very high risk (possible, even very serious, damage to property and persons)***

Here, ***low values*** on the 01–10 scale actually express ***HIGH/REASONABLE detectability.***

Let's look at a couple of examples—see Fig. 4.44 (D-FMEA) and Fig. 4.45 (P-FMEA)—of detection assessments:

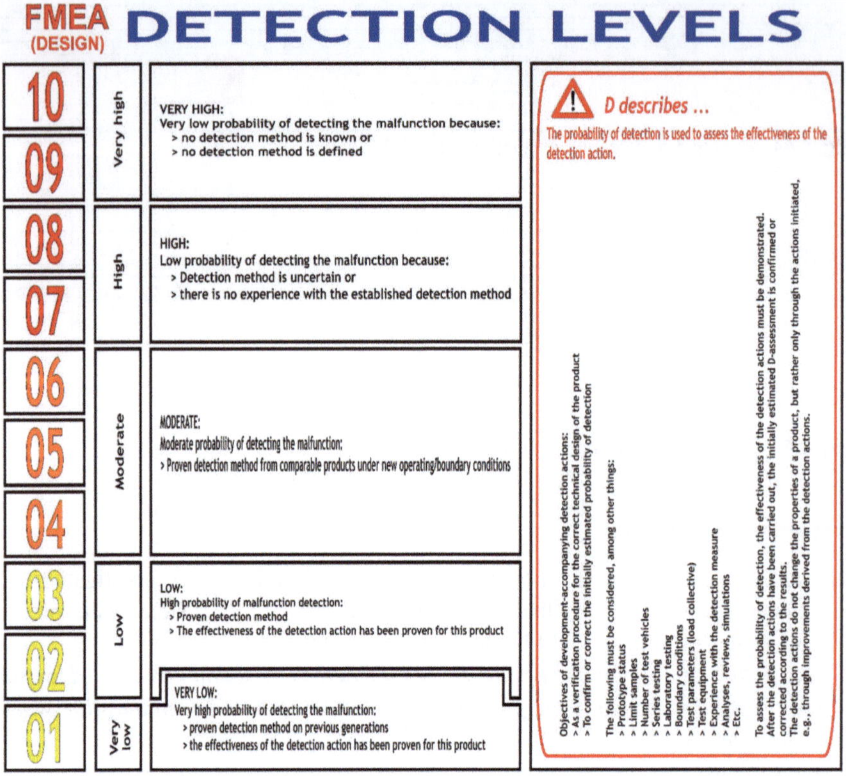

Fig. 4.46 VDA Detection table—D-FMEA

The ***VDA evaluation tables for the detection parameter*** provide the following to support the identification of the detection levels—see Fig. 4.46 (D-FMEA) and Fig. 4.47 (P-FMEA).

Two points should be noted here:

- Columns JL-M have (of course) not changed, as the optimizing criterion (our Lindland's matrix) is a typical S/O matrix criterion. It therefore ***does not*** consider parameter D. This has a very specific meaning in technical FMEA literature, because in fact, ***the only decisive parameter for evaluating the quality of an FMEA from a technical perspective is exclusively the parameter O (referring to the management of failure causes, of course). A company is "better/more capable" the more it PREVENTS, not the more it DETECTS/CONTROLLES***
- The RPN column (Risk Priority Number), however, has assumed its numeric value. Since RPN = (S x O x D), the software automatically calculates the corresponding RPN value as soon as the complete set of values is entered

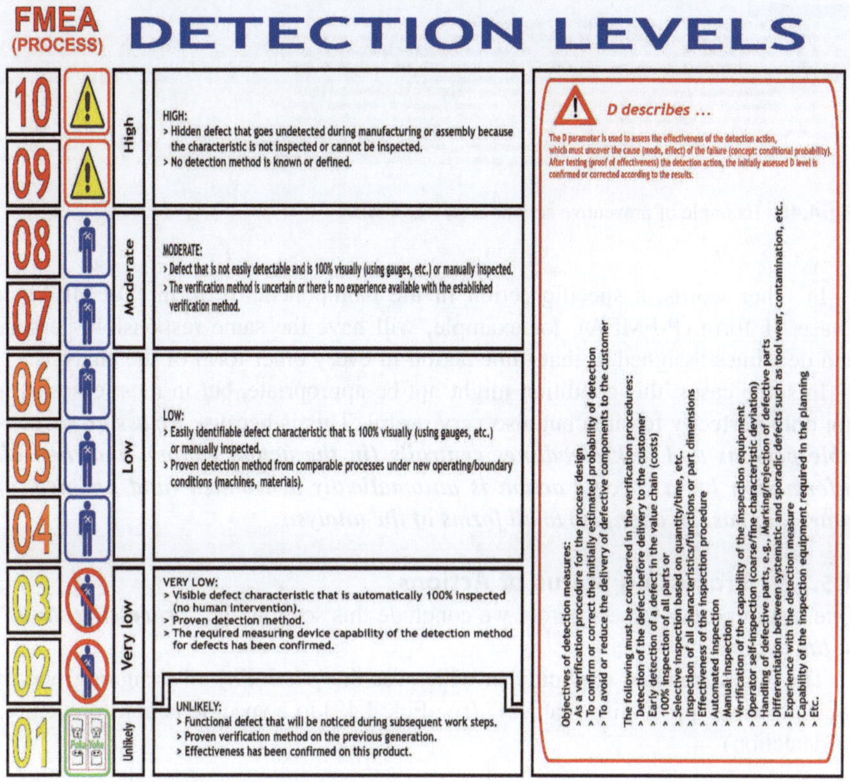

Fig. 4.47 VDA Detection table—P-FMEA

4.5.1.7 Centralized Management of Actions

Definition of "Global" for the Actions

Assigning Responsible Persons to Actions

Assigning Deadlines (Milestones) to Actions

When working with the forms, we define the necessary actions (preventive and detection) line by line (failure cause by failure cause) without immediately worrying about assigning responsibilities and deadlines.

However, once the work in the forms has been completed, we can usefully dedicate ourselves to precisely these two activities (assigning responsibilities and setting deadlines) using the special editor provided by APIS-IQ. Here we see an example (see Fig. 4.48, which considers a D-FMEA) of preventive measures:

First, ***we define the actions as "Global"***.

"Global action" in APIS-IQ means that it has a single responsible person and a single deadline (***global*** means ***in all forms of the relevant FMEA analysis***).

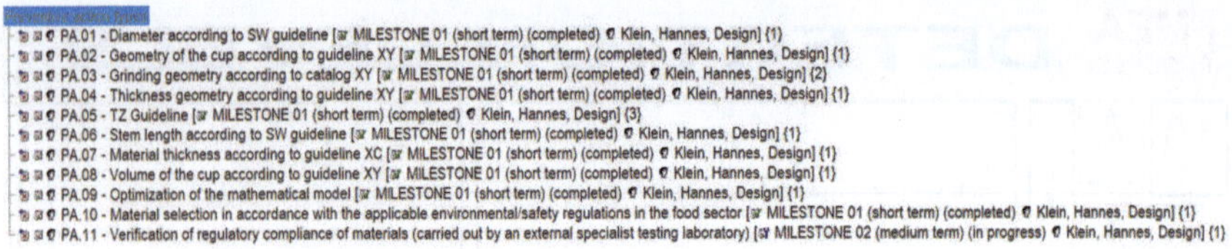

Fig. 4.48 Example of preventive actions (PA)—D-FMEA

In other words, a specific action in the Component 01 form (D-FMEA) or Phase 01 form (P-FMEA), for example, will have the same responsible persons and deadlines assigned to that same action in every other form of the analysis.

In some cases, this condition might not be appropriate, but in most cases, it is not only perfectly feasible but also very useful. This is because ***we assign responsible persons and tasks/deadlines centrally (in the action editor), and then all information for a specific action is automatically distributed (and updated, of course, in case of changes) to all forms of the analysis.***

4.5.1.8 Processing Status of Actions

One last important aspect before we conclude this section: the "***processing status***" ***of the actions.***

The VDA standard explicitly provides for the possibility of assigning one of the following "processing statuses" (see Fig. 4.49) to a specific action (preventive or detection).

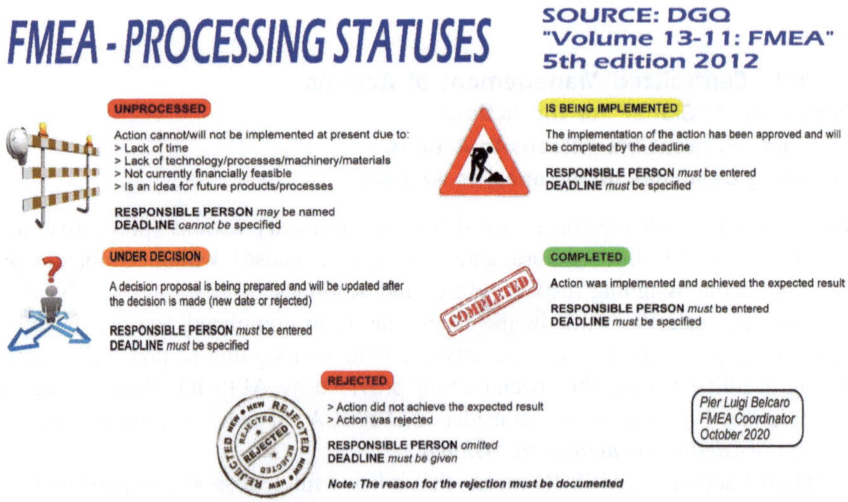

Fig. 4.49 Processing statuses of actions in a D- and P-FMEA

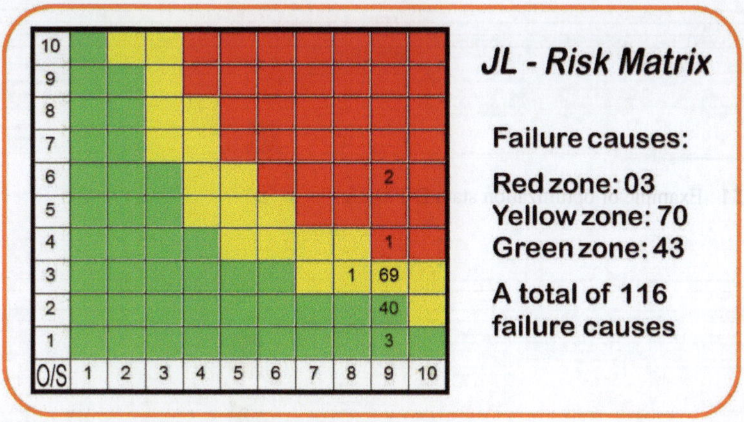

Fig. 4.50 Optimization criterion S/O matrix by John Lindland

4.5.2 LOGICAL BLOCK 03: RISK ANALYSIS, Part 2 (VDA Method: Step 5)

4.5.2.1 Review of the Final Jl Matrix

Managing Points in the Red Zone
Managing Points in the Yellow Zone

Once all the forms of an FMEA have been completed, i.e., after the action analysis and risk assessment (step 4 of the VDA method) have been carried out, ***the fifth and final step (optimization) of the VDA method can be started***.

Optimizing an FMEA Always Requires Defining a Specific Criterion, in This Case, the "Optimization Mechanism" or "Optimization Criterion"

As an optimization criterion, we use John Lindland's S/O risk matrix (as mentioned in previous sections). At the end of an analysis, this could therefore be a typical situation (see Fig. 4.50).

where some failure causes are in the red zone, others in the yellow, and still others in the green zone. In the example above, we see specifically that out of a total of 116 failure causes:

- 3 causes (2.58%) are in the red zone
- 70 causes (60.34%) are in the yellow zone
- 43 causes (37.08%) are in the green zone

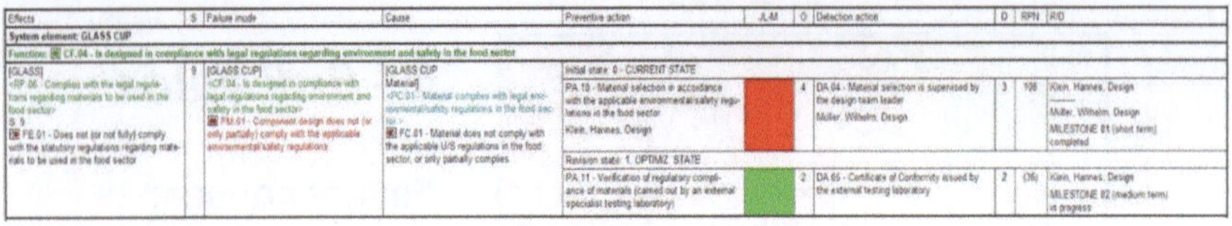

Fig. 4.51 Example of optimization state D-FMEA

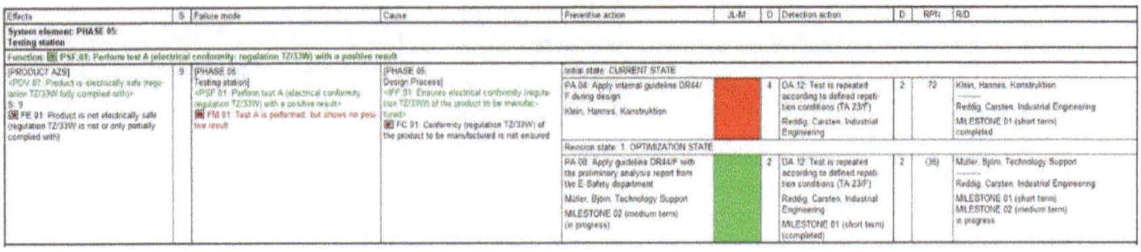

Fig. 4.52 Example of optimization state P-FMEA

Regarding an optimizing risk matrix, the FMEA technical literature specifies the following:

- **RED ZONE (the risk is NOT acceptable)**: ***Points in the red zone MUST be addressed with further and stronger actions***. This means the work team ***MUST*** define improvement actions so that the value of risk parameter O is reduced to such an extent that the original red point is no longer red (i.e., it is in the yellow or green zone)
- **YELLOW ZONE (the risk CAN be considered acceptable)**: ***Points in the yellow zone CAN be addressed with further (stronger) measures***. In this case, it is ***a decision made freely*** by the work team, whereby cost–benefit ratios, time expenditure, resources, etc., should always be carefully weighed
- **GREEN ZONE (the risk IS acceptable)**: ***Points in the green zone do NOT represent a problem*** and therefore require no further treatment

In the example shown above, ***it will therefore be NECESSARY to address three failure causes***. This means that for the three form lines corresponding to the points of interest, ***a new action status (OPTIMIZATION STATE) must be defined***, in which new (and more stringent) preventive actions must be entered.

Let's look at two examples—see Fig. 4.51 (D-FMEA) and Fig. 4.52 (P-FMEA)—for point (S = 9, O = 4).

Here we can see that the original preventive actions:

- ***PA.10***: ***Material selection …*** (D-FMEA)
- ***PA.04***: ***Apply internal guideline DR44/F …*** (P-FMEA)

in the first optimization stage are replaced by the new ***improvement actions***:

- ***PA.11***: ***Verification of regulatory compliance …*** (D-FMEA)
- ***PA.08***: ***Apply guideline DR44/F with the preliminary analysis report …*** (P-FMEA)

which reduces the ***occurrence level*** in both cases ***from an initial value of 4 to a final value of 2.***

Consequently, the form row that corresponded to cell (or point) (S = 9, O = 4) in the Lindland matrix now has the ***new position (S = 9, O = 2)*** in the same matrix and therefore moves from the red to the green zone.

Definition of Priority Criteria (Severity First)

Generation of Specifically Filtered Forms for the Targeted Analysis of Yellow Points

Yellow points should be prioritized, meaning that ***criteria/rules*** should be established to determine which points take precedence over others.

In this case, according to best technical literature, ***parameter S is always considered first***.

Therefore, the following procedure should be followed in the example above:

- 69 points where S = 9, O = 3
- 1 point where S = 8, O = 3

For the 69 points (S = 9, O = 3), a further priority rule could now be established. For example, ***the value of parameter D*** could be considered for each affected failure cause.

Let's assume the work team considers ***all failure causes where the detection D is equal to or greater than 4 to be unacceptable***. This means that ***a subset of the original 69 root causes would be isolated***, requiring improvement. In this case, these could be either preventive (P) or detection (D) actions, depending on the specific technical context.

Finally, ***it can be very useful to specifically filter the forms*** to display only the rows that correspond to yellow cells in the JL matrix. In this context, the filters/advanced filter options of APIS-IQ can be very helpful.

4.5.2.2 Deadline Management

Team Definition of All Action Deadlines
Definition of the Deadlines for the Effectiveness Test
Design and Process Freeze

As we saw above when discussing symbolic deadlines, for operational efficiency reasons associated with centralized data management, it is particularly useful to define the deadlines of the various actions by aligning them as closely as possible with the milestones outlined in the project plan.

Specifically, ***the most concrete and useful approach might be to isolate three deadlines with different time horizons***:

- ***SHORT-TERM*** actions
- ***MEDIUM-TERM*** actions
- ***LONG-TERM*** actions

Each action should then be appropriately assigned to one of these three deadlines. Of course, what exactly "short-, medium-, and long-term" means in the specific context of the FMEA in question is a technical matter for the work team to determine.

Regardless of what the team defines in the previous point, ***it is also necessary to schedule a deadline specifically for reviewing the effectiveness of the actions.***

During the effectiveness review, ***the work team technically assesses, justifies, and documents whether and to what extent the actions taken were actually effective, i.e., whether they achieved the goal (or goals) for which they were originally intended***.

If an action proves ***ineffective***, ***all technical arguments*** for which the team actually considers the action ineffective ***must be documented precisely and in detail***.

Once the effectiveness review is considered complete by the work team (without objections or further comments), the corresponding ***product*** or ***process development*** can be considered "***frozen***" (***design or process freeze***) with regard to the risk analysis.

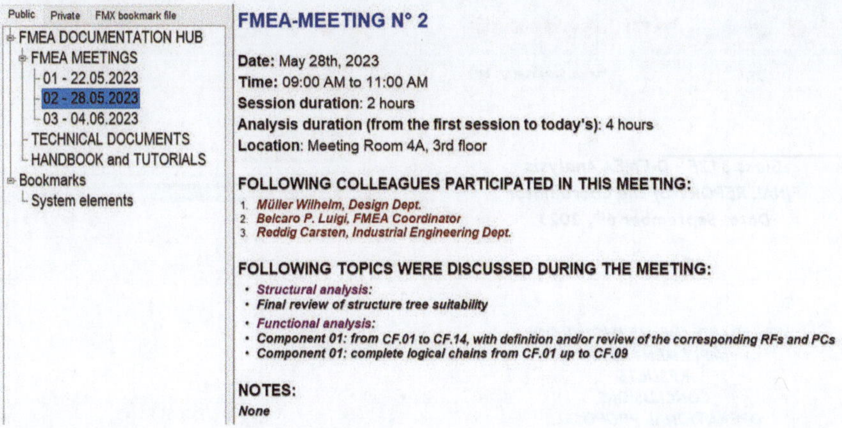

Fig. 4.53 Example of Meeting Log D-FMEA

4.5.2.3 Fmea Documentation

Meeting Records: What and How (Pim Editor)

Final Report and Appendices

Since an ***FMEA is a quality document***, it must also meet the explicitly stipulated requirements for all documents of this class.

APIS-IQ provides the FMEA coordinator with a very useful editor for directly inserting the meeting log into the working file (which contains all the technical data: structure tree, functional and failure frameworks, forms, etc.).

This editor is called PIM (Personal Information Manager) and is typically used to create ***dendritic structures*** (hierarchical trees). Here is a simple example (see Fig. 4.53) for a D-FMEA:

As can be seen from the example, ***the meeting log must contain (at least) the following information***:

- Date
- Time
- Session duration
- Total duration (cumulative time from the first to the current session)
- Location
- Participants
- Topics covered
- Remarks/Comments

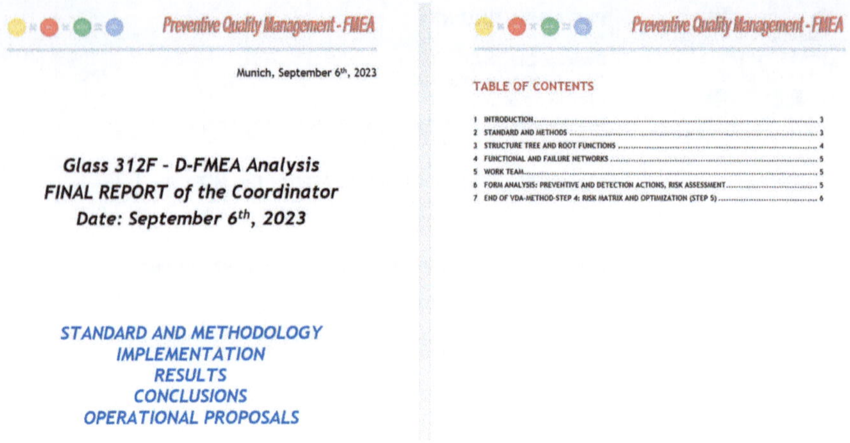

Preventive Quality Management - FMEA

Munich, September 6th, 2023

Glass 312F - D-FMEA Analysis
FINAL REPORT of the Coordinator
Date: September 6th, 2023

STANDARD AND METHODOLOGY
IMPLEMENTATION
RESULTS
CONCLUSIONS
OPERATIONAL PROPOSALS

Preventive Quality Management - FMEA

TABLE OF CONTENTS

Fig. 4.54 Example of Risk Report, pages 1–2 (D-FMEA)

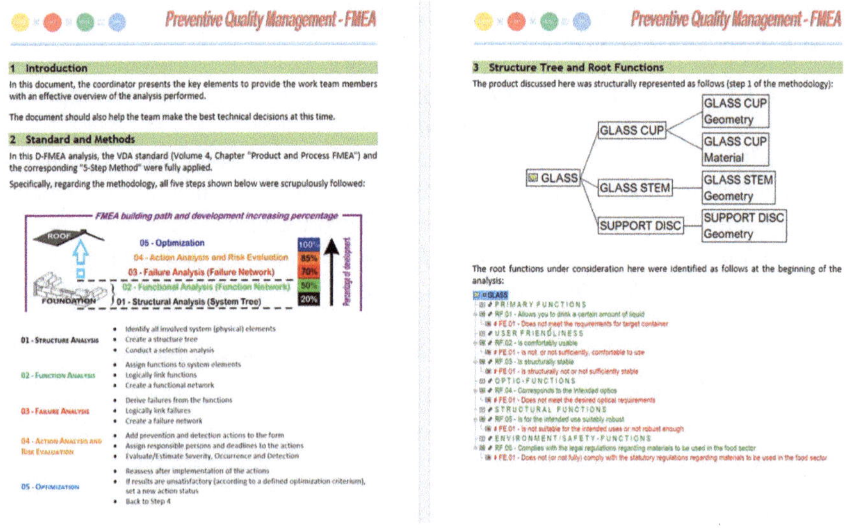

Preventive Quality Management - FMEA

1 Introduction

In this document, the coordinator presents the key elements to provide the work team members with an effective overview of the analysis performed.

The document should also help the team make the best technical decisions at this time.

2 Standard and Methods

In this D-FMEA analysis, the VDA standard (Volume 4, Chapter "Product and Process FMEA") and the corresponding "5-Step Method" were fully applied.

Specifically, regarding the methodology, all five steps shown below were scrupulously followed:

01 - STRUCTURE ANALYSIS
- Identify all involved system (physical) elements
- Create a structure tree
- Conduct a selection analysis

02 - FUNCTION ANALYSIS
- Assign functions to system elements
- Logically link functions
- Create a functional network

03 - FAILURE ANALYSIS
- Derive failures from the functions
- Logically link failures
- Create a failure network

04 - ACTION ANALYSIS AND RISK EVALUATION
- Add prevention and detection actions to the form
- Assign responsible persons and deadlines to the actions
- Evaluate/Estimate Severity, Occurrence and Detection

05 - OPTIMIZATION
- Reassess after implementation of the actions
- If results are unsatisfactory (according to a defined optimization criterium), set a new action status
- Back to Step 4

Preventive Quality Management - FMEA

3 Structure Tree and Root Functions

The product discussed here was structurally represented as follows (step 1 of the methodology):

The root functions under consideration here were identified as follows at the beginning of the analysis:

GLASS
PRIMARY FUNCTIONS
RF.01 - Allows you to drink a certain amount of liquid
FE.01 - Does not meet the requirements for target container
USER FRIENDLINESS
RF.02 - Is comfortably usable
FE.01 - Is not, or not sufficiently, comfortable to use
RF.03 - Is structurally stable
FE.01 - Is structurally not or not sufficiently stable
OPTIC-FUNCTIONS
RF.04 - Corresponds to the intended optics
FE.01 - Does not meet the desired optical requirements
STRUCTURAL FUNCTIONS
RF.05 - Is for the intended use suitably robust
FE.01 - Is not suitable for the intended uses or not robust enough
ENVIRONMENT/SAFETY-FUNCTIONS
RF.06 - Complies with the legal regulations regarding materials to be used in the food sector
FE.01 - Does not (or not fully) comply with the statutory regulations regarding materials to be used in the food sector

Fig. 4.55 Example of Risk Report, pages 3–4 (D-FMEA)

It is also very important and useful for the coordinator to provide a ***final technical* "*Risk Report*"** (typically at the end of the analysis, but I would also recommend it at intermediate stages). Such a final report would generally look like this—see Figs. 4.54, 4.55, 4.56, 4.57 (D-FMEA).

Various supporting documents related to the FMEA can supplement the coordinator's technical risk report, if necessary, as APIS-IQ provides many very useful options in this regard. Some examples may include:

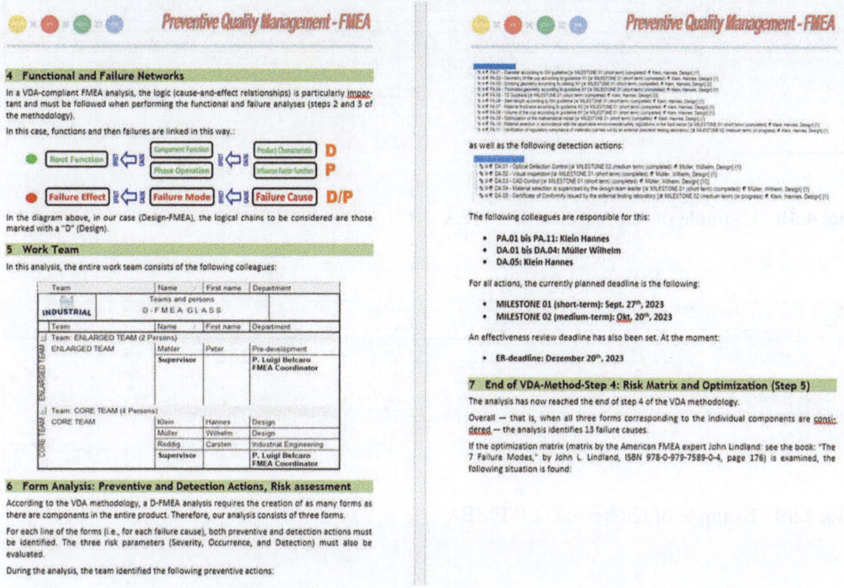

Preventive Quality Management - FMEA

4 Functional and Failure Networks

In a VDA-compliant FMEA analysis, the logic (cause-and-effect relationships) is particularly important and must be followed when performing the functional and failure analyses (steps 2 and 3 of the methodology).

In this case, functions and then failures are linked in this way.:

In the diagram above, in our case (Design-FMEA), the logical chains to be considered are those marked with a "D" (Design).

5 Work Team

In this analysis, the entire work team consists of the following colleagues:

Team	Name	First name	Department
INDUSTRIAL	Teams and persons D-FMEA GLASS		
Team: ENLARGED TEAM (2 Persons)			
ENLARGED TEAM	Mahler	Peter	Pre-development
	Supervisor		**P. Luigi Belcaro FMEA Coordinator**
Team: CORE TEAM (4 Persons)			
CORE TEAM	Klein	Hannes	Design
	Müller	Wilhelm	Design
	Reddig	Carsten	Industrial Engineering
	Supervisor		**P. Luigi Belcaro FMEA Coordinator**

6 Form Analysis: Preventive and Detection Actions, Risk assessment

According to the VDA methodology, a D-FMEA analysis requires the creation of as many forms as there are components in the entire product. Therefore, our analysis consists of three forms.

For each line of the forms (i.e., for each failure cause), both preventive and detection actions must be identified. The three risk parameters (Severity, Occurrence, and Detection) must also be evaluated.

During the analysis, the team identified the following preventive actions:

Preventive Quality Management - FMEA

as well as the following detection actions:

- DA.01 - Optical Detection Control [MILESTONE 02 (medium term) (completed) Müller, Wilhelm, Design] (1)
- DA.02 - Visual inspection [MILESTONE 01 (short term) (completed) Müller, Wilhelm, Design] (1)
- DA.03 - CAD Control [MILESTONE 01 (short term) (completed) Müller, Wilhelm, Design] (1)
- DA.04 - Material selection is supervised by the design team leader [MILESTONE 01 (short term) (completed) Müller, Wilhelm, Design] (1)
- DA.05 - Certificate of Conformity issued by the external testing laboratory [MILESTONE 02 (medium term) (in progress) Klein, Hannes, Design] (1)

The following colleagues are responsible for this:

- **PA.01 bis PA.11: Klein Hannes**
- **DA.01 bis DA.04: Müller Wilhelm**
- **DA.05: Klein Hannes**

For all actions, the currently planned deadline is the following:

- **MILESTONE 01 (short-term): Sept. 27th, 2023**
- **MILESTONE 02 (medium-term): Okt. 20th, 2023**

An effectiveness review deadline has also been set. At the moment:

- **ER-deadline: Dezember 20th, 2023**

7 End of VDA-Method-Step 4: Risk Matrix and Optimization (Step 5)

The analysis has now reached the end of step 4 of the VDA methodology.

Overall — that is, when all three forms corresponding to the individual components are considered — the analysis identifies 13 failure causes.

If the optimization matrix (matrix by the American FMEA expert John Lindland: see the book: "The 7 Failure Modes," by John L. Lindland, ISBN 978-0-979-7589-0-4, page 176) is examined, the following situation is found:

Fig. 4.56 Example of Risk Report, pages 5–6 (D-FMEA)

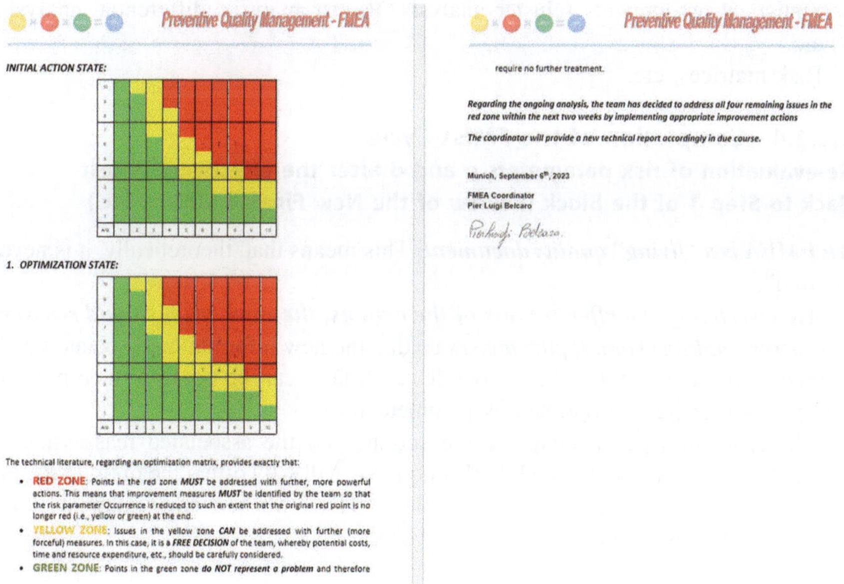

Preventive Quality Management - FMEA

INITIAL ACTION STATE:

1. OPTIMIZATION STATE:

The technical literature, regarding an optimization matrix, provides exactly that:

- RED ZONE: Points in the red zone ***MUST*** be addressed with further, more powerful actions. This means that improvement measures ***MUST*** be identified by the team so that the risk parameter Occurrence is reduced to such an extent that the original red point is no longer red (i.e., yellow or green) at the end.
- YELLOW ZONE: Issues in the yellow zone ***CAN*** be addressed with further (more forceful) measures. In this case, it is a ***FREE DECISION*** of the team, whereby potential costs, time and resource expenditure, etc., should be carefully considered.
- GREEN ZONE: Points in the green zone ***do NOT represent a problem*** and therefore require no further treatment.

Preventive Quality Management - FMEA

Regarding the ongoing analysis, the team has decided to address all four remaining issues in the red zone within the next two weeks by implementing appropriate improvement actions

The coordinator will provide a new technical report accordingly in due course.

Munich, September 6th, 2023

FMEA Coordinator
Pier Luigi Belcaro

Fig. 4.57 Example of Risk Report, pages 7–8 (D-FMEA)

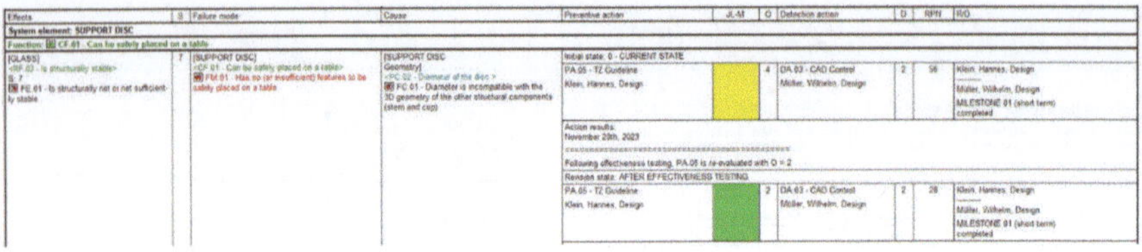

Fig. 4.58 Example of reassessment D-FMEA

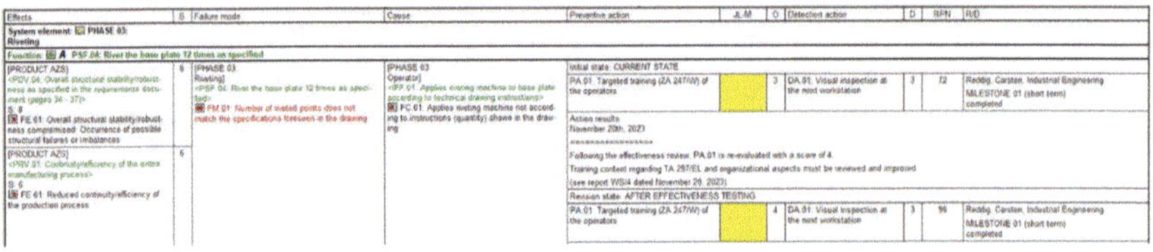

Fig. 4.59 Example of reassessment P-FMEA

- Technical illustrations (e.g., a structure tree, a functional logic chain, etc.)
- Forms (typically in PDF format)
- Forms with targeted filters (for prioritizing actions, for example)
- Statistical assessments (cluster analysis, Pareto analysis, differential analysis, etc.)
- Risk matrices, etc.

4.5.2.4 Completion of the FMEA Cycle

Re-evaluation of risk parameters o and d after the effectiveness test
Back to Step 1 of the Block (Review of the New Final JI Matrix, etc.)

An FMEA is a "living" quality document. This means that, theoretically, it is never "closed".

After verifying the effectiveness of the actions, the work team should reassess the corresponding O and D parameters so that the new information associated with implementing the actions, the lessons learned, etc., can be possibly incorporated into new and more appropriate risk parameters.

All technical aspects related to the actions and the associated reassessments should be carefully documented. In this context, APIS-IQ offers the option "***Result of Actions***" in the form's display options. Here we see two examples—see Fig. 4.58 (D-FMEA) and Fig. 4.59 (P-FMEA).

Naturally, ***a parametric re-evaluation usually also produces a new final JL matrix with new color combinations***.

This means that operationally we should return to step 1 of the block and repeat what is necessary.

The Product/Process Interface

5

Abstract

This chapter introduces the 4 basic types of FMEA, using a representation model of the logical relationships between them according to the staircase scheme of the VDA standard to explain how to correctly handle the corresponding FMEA structures.

5.1 The 4 Basic Types of FMEA: System FMEA, Design FMEA, Process FMEA and Machine FMEA

The current state of the art provides for ***4 basic types*** of risk analysis:

- ***SYSTEM FMEA***: This is an FMEA that is essentially performed "at the macro level." A typical example in the automotive sector is the ***SYSTEMIC analysis of a vehicle***. In this case, we deal at the structural layer with
  - A first node on the left for the ***entire vehicle***
  - Intermediate nodes for the ***various systems*** that make it up
  - Third-level nodes for the ***(most relevant) components*** of the aforementioned systems
- ***DESIGN FMEA***: A ***"Design FMEA" (or "Product FMEA")*** is an analysis in which we instead find—always with reference to the first, i.e., the structural layer—(from left to right):
  - (Entire) ***Product***
  - ***Product Components***
  - ***Technical specifications*** (of the components)
- ***PROCESS FMEA***: A ***Process FMEA*** has a structural layer as follows:
  - ***Product*** (as the ***target of*** the corresponding industrial ***production process***)
  - ***Phases*** of the production process

P. L. Belcaro, *9-Object-Model FMEA*, https://doi.org/10.1007/978-3-658-50832-6_5

- *Influence factors*: the so-called "***5 M*** (German alphabet) ***rule***". In other words:
  Man (Human Factor)
  Material
  Machine
  Environment
  Method (Methods/Processes)
- ***MACHINE FMEA:*** A ***Machine FMEA*** is essentially the ***D-FMEA of a machine/device*** used in the production process.

5.2 Conceptual Relationship According to the VDA Standard Between the Different Types of FMEA: The "Staircase Scheme"

The German standard links the FMEA types according to very precise ***logical relationships*** (see Fig. 5.1), which can be used particularly effectively when a risk analysis for more than one FMEA type is required.

In the image above, the abbreviations used to mark the three steps each "layer" consists of (the "landings" of the staircase) have the functional (green circles) and dysfunctional (red circles) meanings that we are already well familiar with.

We can already deduce from the proposed diagram that ***the logical relationship between the different layers (landings, or the different FMEA types) is based on correspondences shifted by one step***. For example:

- A failure mode (FM) in a Design FMEA (D-FMEA) must find a relevant correspondence (here in the sense of specific technical content) in a failure effect (FE) of the corresponding Process FMEA (P-FMEA)
- In other words, if we now shift the discussion to the hierarchically higher analytical layer, i.e., the functional layer, then in a P-FMEA, the elements above referred to as "product variables" must find a solid content background in just as many "component functions" (D-FMEA)

This ***rational pattern*** is replicated analogously for all other "landings" described in the staircase diagram.

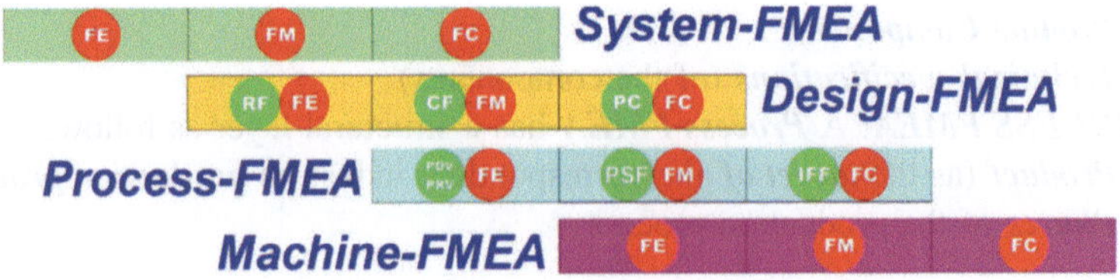

Fig. 5.1 The VDA "Staircase Scheme"

5.3 What is Particularly Important to Ensure in the Design/Process Information Exchange and Why

5.3.1 Critical Geometries

5.3.2 Critical Material, Physical and Chemical Properties

5.3.3 Actual Ability of the Industrial Production Process to Realize Technical Specs as Defined in the Design (Technical Drawings)

The industrialization of a product is generally never a small undertaking, a "minor activity", we could say.

In fact, it is usually an activity that, at least in the initial phase, requires a significant investment of time and resources and necessitates the presence of several solid technical skills and good coordination between the various technical departments.

The ***starting point*** is always the ***company's design function*** (even before the R&D—"Research & Development" function—but we will not discuss this aspect here, although it is absolutely relevant).

Based on customer specifications (customer requirements specification) and generally other specifications from various sources, company designers develop a specific product and summarize the majority of their work in one or more technical drawings (in addition to the various CAD 3D files created).

The responsibility then passes from the designer to the Industrial Engineering (or Industrial Production Process Planning) department. Here, the focus is on all technical aspects related to the ***industrial feasibility/manufacturability of the product*** itself. This primarily involves a ***precise and detailed exchange of information between the designer and the industrial engineer***.

Typically, the ***optimal overlap*** between the ***product's technical specifications***, as defined in the design (technical drawing), and the ***materials, machinery, equipment, lines,*** **etc.**, ***related to the production process*** is deeply analyzed. Only in this way there can be a reasonable assurance that the product in question.

- ***can actually be manufactured***, given the various industrial process factors available to the company
- can also be manufactured in accordance with the most comprehensive and robust ***technical efficiency criteria*** (minimizing industrial costs, minimizing resource use, limiting technical times, etc.)

One aspect of absolute importance that must be emphasized here is that ***FMEA, viewed as an "internal semantic code" within the company, can be of enormous benefit for optimal information exchange between the various technical parties***.

However, one key aspect must be given particular attention: it is absolutely essential that the FMEA coordinator takes very good care of the following:

- First, the structuring of the various information exchange activities (logical working frame)
- and then the development of the entire operational procedure and all related tools to support these activities (typically well-defined software procedures)

If all the above-described prerequisites are met, then FMEA can fulfill its ideal, highly advantageous function as a ***corporate glue***—a function that any enlightened management team would want to see consistently and widely implemented within their own organization.

Regarding the specific dimensions/variables to be considered, the key aspects in this particular context are essentially the following:

- The ***critical geometries***
- The ***critical material properties***
- The ***critical physical and chemical properties of the product***

The aforementioned design variables ***must be effectively and efficiently reflected in the actual technical capability of the industrial production process***. This is essentially the core of the statement.

5.4 Design Conceptual Model (Structure)

5.4.1 D.01: Product

5.4.2 D.02: Components

5.4.3 D.03: Technical Specifications (of the Components)

Let us now look at a concrete example of how an ***appropriate product/process interface*** can be implemented. Let us assume that our product (see Fig. 5.2) is the following.

This is a typical component for the ***aerodynamic airflow around a vehicle's windshield*** and also serves structural functions (housing the wiper pins) and facilitates rainwater drainage.

We will now consider the following functional chain (see Fig. 5.3).

Here we analyze the ***root function RF.01***, which has its logical basis (cause/effect) in the ***component function CF.01***. This component function, in turn, is logically explained by ***two functional explainers, PC.01 and PC.02***, at the level of the technical specifications of component "Part 1" (***PC*** stands for ***product characteristic***, as we already know).

We now turn to the production process.

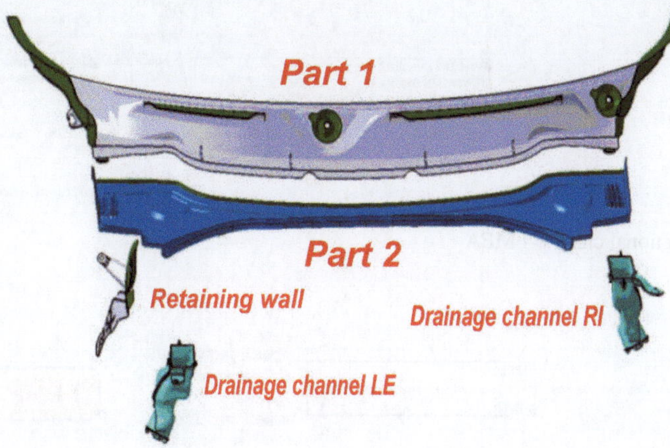

Fig. 5.2 Example part "Windscreen"

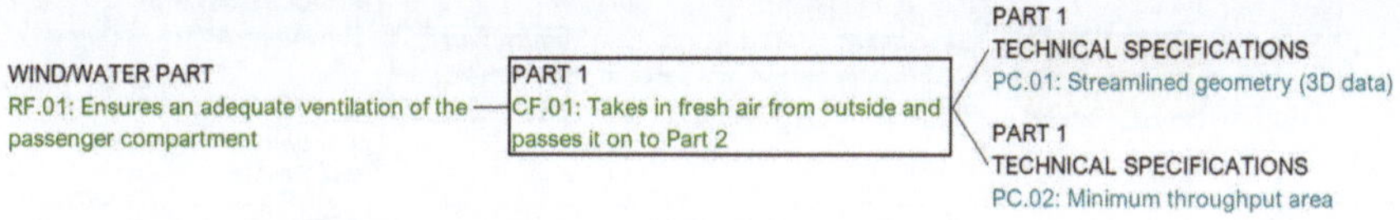

Fig. 5.3 Functional chain D-FMEA

5.5 Production Process Conceptual Model (Structure)

5.5.1 P.01: Target Product of the Production Process

5.5.2 P.02: Process Phases

5.5.3 P.03: Influence Factors ("5 m Rule")

At the level of the production process, we could have a functional chain (see Fig. 5.4) of this type.

Let us now compare the two chains (design and process) and make some considerations using the reference logic diagram (staircase diagram)—see Fig. 5.5

Component function CF.01 results in ***product variable PDV.01***.

The relevant technical issue here is that ***the production process is capable of delivering a product that fully meets the "air suction capability" characteristics originally defined by the designer***.

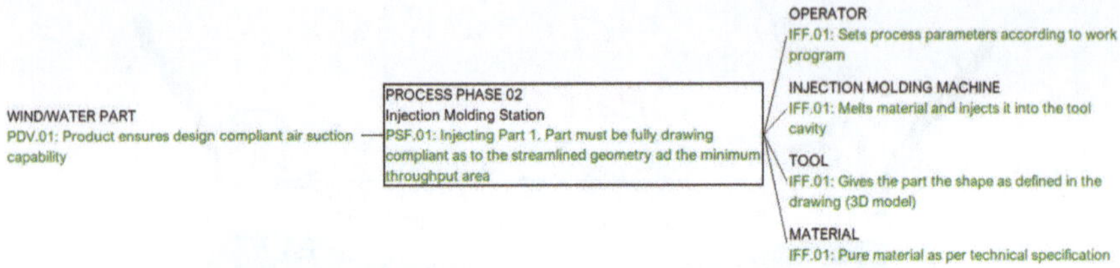

Fig. 5.4 Functional chain P-FMEA

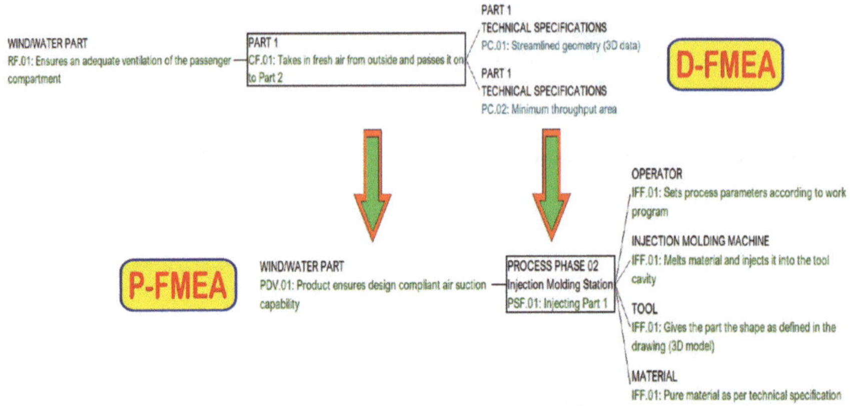

Fig. 5.5 Comparison of functional chains (D- and P-FMEA)

5.6 D/P Interface

5.6.1 D.02 → P.01 (Identification of Relevant Product Variables)

5.6.2 D.03 → P.02 (Design-Oriented Definition of Phase Operations)

As soon as we move from CF.01 one "step" to the right (to the technical specifications) at the "D-FMEA landing" (staircase scheme), we can immediately understand what ***air intake capability*** means technically:

- ***Streamlined geometry***
- ***Minimum throughput area***

Therefore, at the production level, the PSF.01 of Phase 02 (injection molding station) must absolutely take these two ***design constraints*** into account. So much so that the PSF.01 should correctly be reformulated as follows (see Fig. 5.6).

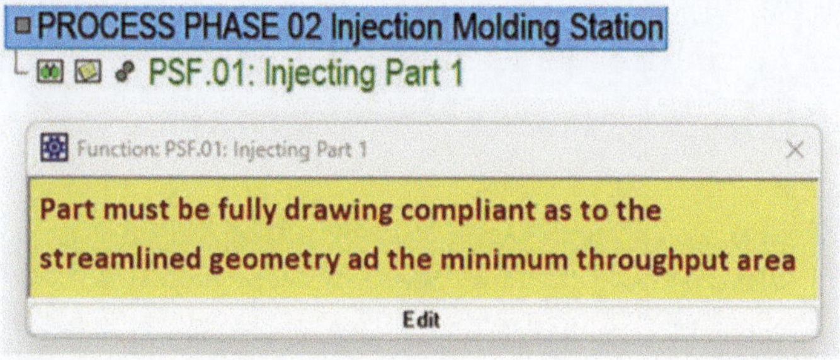

Fig. 5.6 PSF.01 taking into account the design constraints

From this point on, the Industrial Engineer has all conceivable options. How? By acting on the basis of their technical industrialization expertise and their experience with the company resources and tools available for industrial production.

In particular, the ***IE will define the most suitable set of possible influencing factors*** to logically (cause/effect) fulfill and justify the phase operation (Phase Step Function: PSF.01). In this case, we see that the IE has four basic "levers" at their disposal:

- Setup technician
- Injection molding machine
- Tool
- Material

The influence factors define the narrowest scope of the IE, just as the narrowest scope of the designer is reserved for technical specifications.

A "***winning team***" is precisely that in which ***designers and IEs best perform their specific technical functions, particularly at the third*** (vertical, see Fig. 2.3: Working Model—Layers and Levels) ***level of their respective functional layer***:

- Technical specifications for the designer
- Influence factors for the IE

The ***FMEA coordinator*** is responsible, in any case, ***for structuring the work of the designer and IE methodically and procedurally***.

Essentially, ***a good coordinator provides his/her main technical interlocutors (designer and IE) with the best possible reference frame*** so that they can integrate their best technical content into the most effective and efficient conceptual/methodological framework.

In other words, we could perhaps also say metaphorically that the coordinator provides a masterfully painted diptych with precisely the frame that best highlights the beauty of the two panels (the technical content of design and process).

Advanced Tools 6

Abstract

This chapter presents a mathematical model for evaluating Occurrence and Detection risk parameters, based on specifically formulated vector indicators instead of conventional rating tables.

6.1 A Mathematical Model for Risk Assessment: Introduction

This section of the book aims to introduce a new approach to the risk evaluation phase of an FMEA. In this case, the outlined approach is mathematical in nature. Applied mathematics can be very helpful in refining decision-making techniques for assigning values to risk parameters in an FMEA.

Every FMEA coordinator has undoubtedly experienced varying degrees of uncertainty in their experience when trying to assign unambiguous values to risk parameters based on classic rating tables (AIAG or VDA). This is due to the ***pronounced subjective nature of the various selection options***. For example, what does it mean that the occurrence of a specific failure cause (given a certain preventive action) is rated as 5? Let's take a look at the classic tables.

According to the ***AIAG standard*** (see Fig. 6.1), an occurrence of 5 is based on the following information:

According to the ***VDA standard***, however, we have 5 if the following information (see Fig. 6.2) applies:

As we can easily imagine, it's not simple to say that a particular occurrence is a 5 for both me and you, since I might select a 5 based on the guidance suggested in the table, while you might select a 4 (for example) based on the same information. The ***degree of subjectivity*** in this case is anything but negligible. This poses a real problem, as the overall FMEA assessment could then differ significantly depending on whether it's conducted by me or by you.

P. L. Belcaro, *9-Object-Model FMEA*, https://doi.org/10.1007/978-3-658-50832-6_6

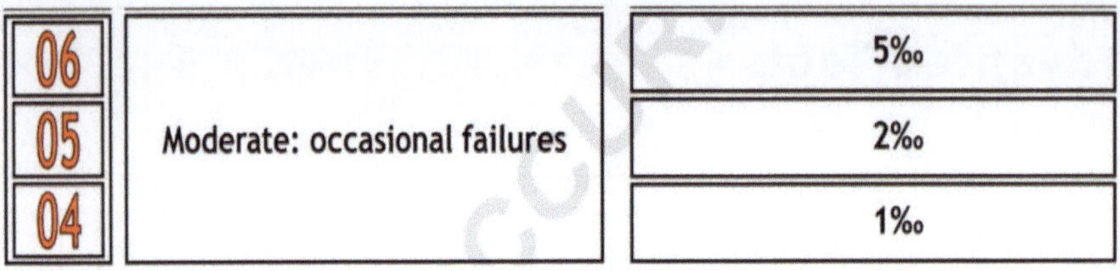

Fig. 6.1 Assessment table for occurrence (AIAG standard, D-FMEA)

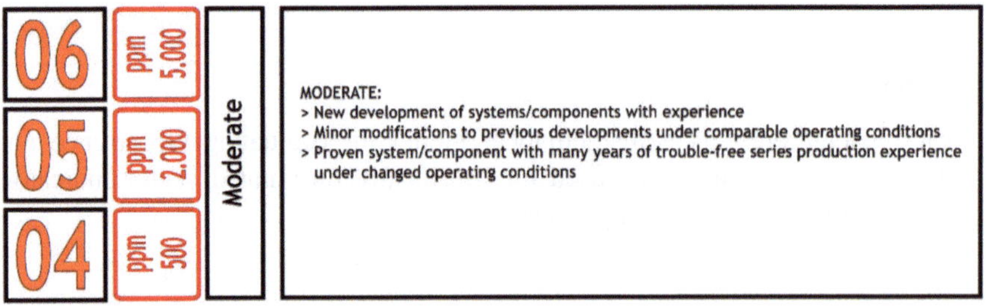

Fig. 6.2 Assessment table for occurrence (VDA standard, D-FMEA)

It is well known that ***an FMEA inherently contains a physiological element of subjectivity***. However, this subjectivity should be limited—as far as possible—by "suitable rules."

Is it possible to define "suitable rules"? I think so. For this reason, I wanted to employ applied mathematics and formulate a correspondingly new approach. At the end of this section, we will then try to understand together the strengths and weaknesses of this innovative approach.

6.2 Mathematical Evaluation of Occurrence

6.2.1 Input Parameters (Occurrence Explainers)

The basic idea is to first ***define a set of reasonable parameters***. These parameters are then used ***to explain the levels of occurrence***. Let's assume we have ***4 parameters*** with which we want to express the level of occurrence of a specific failure cause (given a certain preventive action):

	3	2	1
Technical manageability of the failure cause	Cause is difficult to be managed by design	Cause is manageable with some effort	Cause is simply/effectively manageable by design
Quality history of the failure cause	Cause is new/not documented before	Cause is documented but problematic	Cause is well documented/well known
Strength of the preventive action	Action is weak	Action is helpful enough	Action is strong
Technical feasibility of the preventive action	Action is technically difficult to be implemented	Action can be technically implemented but implementation is problematic	Action is technically easy to be implemented

Fig. 6.3 States (qualitative and quantitative) of the 4 parameters (occurrence explainers)

y_1:	**Technical manageability**	of the **Failure cause**
y_2 :	**Quality history**	
y_3 :	**Strength**	of the **Preventive action**
y_4 :	**Technical feasibility**	

Each parameter y_i has ***3 possible states***. The states are ***both qualitative (descriptions) and quantitative (from 1 to 3)***. Regarding the quantitative states, ***1 means "Good" and 3 means "Bad"*** (see Fig. 6.3).

In mathematical terms, we can determine:

$$y_i \in \{1, 2, 3\} \subseteq \mathbb{N}$$

In this way, each occurrence level now becomes ***a vector of*** $\mathbb{N}^4$:

$$\vec{z} = (y_1, y_2, y_3, y_4) \text{ where : } y_i \in \{1, 2, 3\} \subseteq \mathbb{N}$$

6.2.2 Metrics (Condensation of Information)

The next step is all about to understand ***how to extract a single value that expresses our level of occurrence from the vectors***. Several different selection options exist. We will pursue a simple approach here, knowing that this may not be the optimal choice.

We introduce ***metrics (scalar measurements)*** into the vector space defined above. As the first metric, we consider a ***simple linear function (mean)***.

$$\mu(\vec{z}) = \mu_{i=1}^{4}(y_i) = \frac{1}{4}\sum_{i=1}^{4} y_i$$

We can now easily identify two basic points:

$$\vec{\alpha} = (1, 1, 1, 1) \rightarrow \mu(\vec{\alpha}) = 1 \text{ Best situation}$$

$$\vec{\beta} = (3, 3, 3, 3) \rightarrow \mu\left(\vec{\beta}\right) = 3 \text{ Worst situation}$$

These two points simultaneously define the ***minimum*** and ***maximum*** values of the ***function's codomain***. That is, all values of the average $\mu(\vec{z})$ lie between 1 (minimum) and 3 (maximum).

Now let's consider the following indicator:

$$M(\vec{z}) = Max(\vec{z}) = Max_i(y_i)$$

For the ***second distance*** (metric), we will assume the following expression (vector function):

$$\tau(\vec{z}) = \frac{\sqrt{(M-1)^2 + (\mu - 1)^2}}{2\sqrt{2}}$$

In this second case we find that:

$$\vec{\alpha} = (1, 1, 1, 1) \to \tau(\vec{\alpha}) = 0 \ \text{ Best situation}$$

$$\vec{\beta} = (3, 3, 3, 3) \to \tau\left(\vec{\beta}\right) = 1 \ \text{Worst situation}$$

In this case as well, these two points simultaneously define the minimum and maximum of the function's codomain. This second choice is linked to the following Cartesian observations:

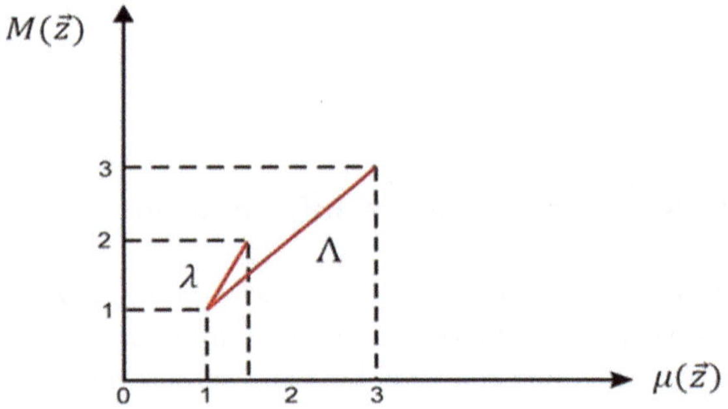

$$\lambda = \sqrt{(M-1)^2 + (\mu - 1)^2}$$

$$\Lambda = 2\sqrt{2}$$

where (1, 1) is the point of lowest risk and (3, 3) is the point of highest risk. We therefore have:

$$\tau(\vec{z}) = \frac{\lambda}{\Lambda} = \frac{\sqrt{(M-1)^2 + (\mu - 1)^2}}{2\sqrt{2}}$$

	3	2	1
Technical manageability of the failure cause	Cause is difficult to be managed by design	Cause is manageable with some effort	Cause is simply/effectively manageable by design
Quality history of the failure cause	Cause is new/not documented before	Cause is documented but problematic	Cause is well documented/well known
Strength of the preventive action	Action is weak	Action is helpful enough	Action is strong
Technical feasibility of the preventive action	Action is technically difficult to be implemented	Action can be technically implemented but implementation is problematic	Action is technically easy to be implemented

Fig. 6.4 Example of parameter individual configuration

6.2.3 Example

Let's look at an example. If we had the following configuration of parameters (see Fig. 6.4, orange color).

we would have the following vector:

$$\vec{s} = (y_1, y_2, y_3, y_4) = (2, 1, 1, 2)$$

The relevant metrics (occurrence level) would then be:

$$\mu(\vec{s}) = \mu_{i=1}^{4}(y_i) = \frac{1}{4}\sum_{i=1}^{4} y_i = \frac{3}{2} = 1.50$$

$$\tau(\vec{s}) = \frac{\sqrt{(M-1)^2 + (\mu - 1)^2}}{2\sqrt{2}} = 0.40$$

6.2.4 Transformation (Linear)

To ensure compatibility with the usual occurrence measurements, which range from 1 to 10, we need to transform the range of the above-mentioned metric functions into the usual range of 1–10.

Our goal now is therefore to bring about a ***transformation*** ’ which is able to meet the following conditions:

$$\left[\mu(\vec{\alpha}), \mu\left(\vec{\beta}\right)\right] \overset{\varphi}{\Rightarrow} [1, 10]$$

$$\left[\tau(\vec{\alpha}), \tau\left(\vec{\beta}\right)\right] \overset{\varphi}{\Rightarrow} [1, 10]$$

In this case, we can simply apply the ***linear transformation***:

$$\omega = \varphi\left[d(\vec{z})\right] = md(\vec{z}) + q$$

where:

$$d(\vec{z}) = \mu(\vec{z}) \text{ or } \tau(\vec{z})$$

Below we give a Cartesian representation for the first of the two assumed metrics:

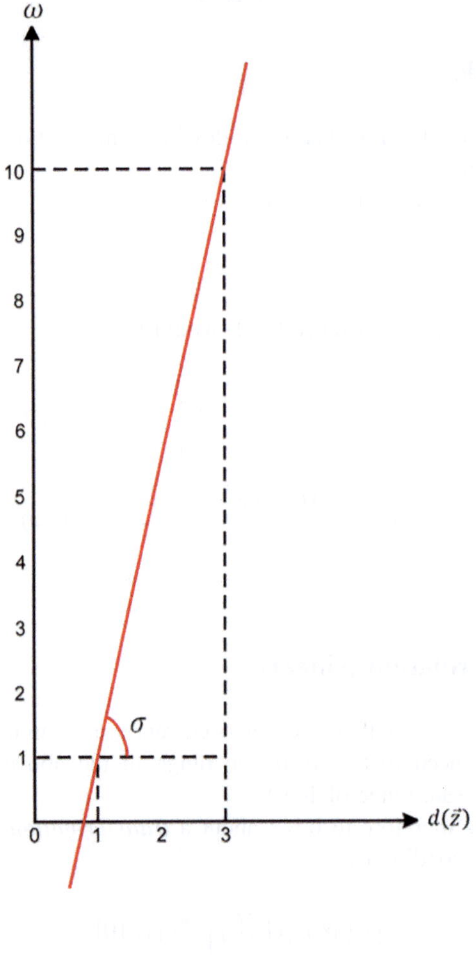

where:

$$m = \tan\sigma = \frac{\sin\sigma}{\cos\sigma} = \frac{9}{2}$$

and

$$q = -\frac{7}{2}$$

For the second metric, we can easily find that:

$$m = 9$$

and

$$q = 1$$

In this way we have obtained a ***linear transformation that can map all vectors expressing a specific level of occurrence into the range 1–10***, as is usually the case in classical rating tables.

6.2.5 Balancing the Information (Weighted Sum)

The first and second transformation could now be ***blended in a "balanced" way***, considering that the first transformation will always produce smoother output values than the second one.

For example, we could assume that we choose a ***final output*** as follows:

$$\omega_{out} = w_1\varphi[\mu(\vec{z})] + w_2\varphi[\tau(\vec{z})]$$

where:

$$w_1 = 40\% \; w_2 = 60\%$$

6.3 Classical (Tabular) and Mathematical Assessment: A Comparison

The difference (and the advantage) of such an approach should be that by adopting this new perspective, we (consistently) limit the subjectivity of a "classic" evaluation. It is much simpler and more effective to define a particular level of occurrence when it is viewed as an expression of defined states across a set of explainers.

It should be noted that ***the explainers selected here are simply based on my personal choice***. Many different choices would be possible, of course. I would suggest analyzing the most suitable set of explainers for a company together with colleagues who have the most valuable experience in design (or production process, for P-FMEA) matters.

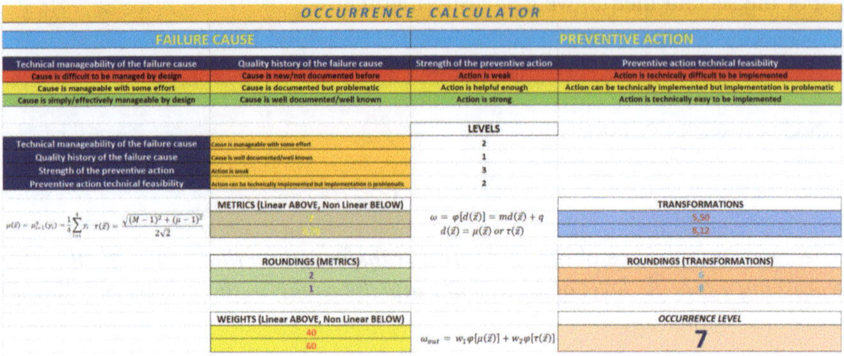

Fig. 6.5 Example of assessment calculator

6.4 Assessment Calculator

The next step is to set up a so-called ***(assessment) calculator***. Using a simple Excel spreadsheet and inserting the appropriate mathematical formulas, we can create an "***smart interface***" that allows us to express a level of occurrence by selecting the appropriate states for each component of the vector $\vec{z}$.

Below is an example (see Fig. 6.5) of what such a calculator might look like.

6.5 Summary of the Procedure

- Given a specific failure cause (and its associated preventive action), whose occurrence we wish to assess, we select values (corresponding to simple and comprehensible qualitative descriptions) from ***4 parameters (explainers)*** quantitatively ranging from 1 (good) to 3 (bad)
- We summarize the information in ***metrics***, which in this case are assumed to be the ***average value*** and a ***biased average value***
- ***We transform*** the metrics from the ranges 1–3 and 0–1 into the classic range 1–10 by applying a simple linear transformation
- ***We merge*** the measurements in a balanced way (by selecting proper weights w)
- This final step allows us to obtain an ***OCCURRENCE LEVEL***.

6.6 Example

Let us consider the same example as before (see Fig. 6.4). Expressed quantitatively, we have:

$$\vec{s} = (y_1, y_2, y_3, y_4) = (2, 1, 1, 2)$$

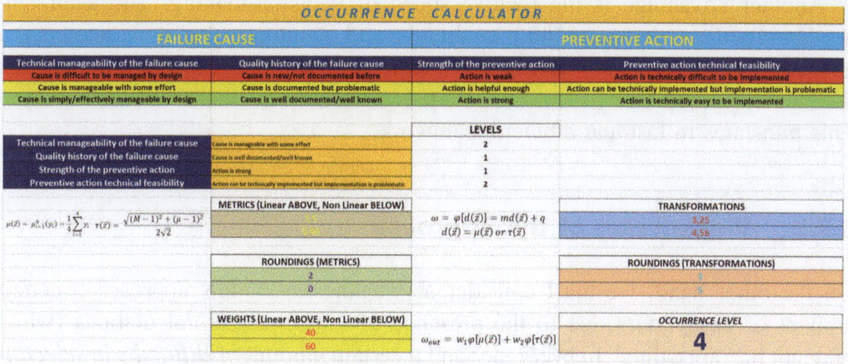

Fig. 6.6 Occurrence assessment calculator

The metrics will be the following:

$$\mu(\vec{s}) = \mu_{i=1}^{4}(y_i) = \frac{1}{4}\sum_{i=1}^{4} y_i = \frac{3}{2} = 1.50$$

$$\tau(\vec{s}) = \frac{\sqrt{(2-1)^2 + (1.50-1)^2}}{2\sqrt{2}} = 0.40$$

And the final ***occurrence level*** will be:

$$\omega_{out} = 0.40\varphi[\mu(\vec{s})] + 0.60\varphi[\tau(\vec{s})] = 4$$

The ***calculator*** helps us to automatically obtain the result (see Fig. 6.6) after we have simply entered the qualitative states:

6.7 Conclusions

To conclude the description of this new mathematical approach, we will now attempt to identify its ***strengths*** and ***weaknesses***. We want to understand under what conditions it might be a better choice to evaluate mathematically rather than in the "classical" way.

6.7.1 Strengths of the Approach

If the risk analysis involves a complex product requiring precision, the traditional approach might not be the best choice, as the assessment of occurrence will inevitably involve a degree of subjectivity. However, if our goal is to minimize subjectivity and emphasize reasonable ***repeatability*** and precision, then this approach could indeed be a good option.

Another advantage is the relative speed of decision-making. The team only needs to focus on the states (both qualitative and quantitative) of 4 parameters that appear very clear and very easy to select. Once the parameters are well-defined and the metrics appropriately identified, the process highly facilitates decision-making. This translates to fast and efficient teamwork.

6.7.2 Weaknesses of the Approach

On the other hand, we must consider as weaknesses of this approach a certain amount of ***complexity*** due to the presence of mathematical calculations (which may not be "appealing" to everyone) and a certain amount of ***difficulty in selecting suitable metrics***.

6.8 Appendix 01: Main Features of The Metrics

It might be interesting to analyze the main characteristics of the metrics involved here. We have here ***4-component vectors (domain)*** and ***scalars (codomain)***. It's easy to see that we have a total of:

$$\mathbf{3^4 = 81}$$

vector points. If we were to consider how these are distributed, we would obtain the following table (see Figs. 6.7, 6.8 and 6.9).

Here, ***M01*** and ***M02*** are the first and second metrics described here, respectively, and ***O*** (Occurrence) is the ***final occurrence level***.

The ***occurrence diagram*** (vector points on the x-axis, corresponding occurrence level on the y-axis) would look like this (see Fig. 6.10).

If we were to consider the ***frequencies of the occurrence levels***, i.e., how often a particular occurrence level is represented in a metric, we would find the following distributions (see Fig. 6.11) (***mean*** on the left, ***mean with bias*** on the right).

In the first case, we have ***a typical bimodal dynamic***, with levels 4, 6 and 7 being the most concentrated.

In the second case, level 7 is the most concentrated. This also helps us understand that considering a balanced mix of the two metrics might be a better approach to obtain more consistent final values.

In any case, selecting a suitable metric or metric mix is not easy, as multiple variables may be involved, each of which should be correctly analyzed and interpreted.

					MAX	MEAN	TAU	M 01	M 02	O
1	1	1	1	1	1	1	0,00	1	1	1
2	1	1	1	2	2	1,25	0,36	2	4	3
3	1	1	1	3	3	1,5	0,73	3	8	6
4	1	1	2	1	2	1,25	0,36	2	4	3
5	1	1	2	2	2	1,5	0,40	3	5	4
6	1	1	2	3	3	1,75	0,76	4	8	6
7	1	1	3	1	3	1,5	0,73	3	8	6
8	1	1	3	2	3	1,75	0,76	4	8	6
9	1	1	3	3	3	2	0,79	6	8	7
10	1	2	1	1	2	1,25	0,36	2	4	3
11	1	2	1	2	2	1,5	0,40	3	5	4
12	1	2	1	3	3	1,75	0,76	4	8	6
13	1	2	2	1	2	1,5	0,40	3	5	4
14	1	2	2	2	2	1,75	0,44	4	5	5
15	1	2	2	3	3	2	0,79	6	8	7
16	1	2	3	1	3	1,75	0,76	4	8	6
17	1	2	3	2	3	2	0,79	6	8	7
18	1	2	3	3	3	2,25	0,83	7	9	8
19	1	3	1	1	3	1,5	0,73	3	8	6
20	1	3	1	2	3	1,75	0,76	4	8	6
21	1	3	1	3	3	2	0,79	6	8	7
22	1	3	2	1	3	1,75	0,76	4	8	6
23	1	3	2	2	3	2	0,79	6	8	7
24	1	3	2	3	3	2,25	0,83	7	9	8
25	1	3	3	1	3	2	0,79	6	8	7
26	1	3	3	2	3	2,25	0,83	7	9	8
27	1	3	3	3	3	2,5	0,88	8	9	9

Fig. 6.7 Vectors (domain) and corresponding scalars (codomain)

					MAX	MEAN	TAU	M 01	M 02	O
28	2	1	1	1	2	1,25	0,36	2	4	3
29	2	1	1	2	2	1,5	0,40	3	5	4
30	2	1	1	3	3	1,75	0,76	4	8	6
31	2	1	2	1	2	1,5	0,40	3	5	4
32	2	1	2	2	2	1,75	0,44	4	5	5
33	2	1	2	3	3	2	0,79	6	8	7
34	2	1	3	1	3	1,75	0,76	4	8	6
35	2	1	3	2	3	2	0,79	6	8	7
36	2	1	3	3	3	2,25	0,83	7	9	8
37	2	2	1	1	2	1,5	0,40	3	5	4
38	2	2	1	2	2	1,75	0,44	4	5	5
39	2	2	1	3	3	2	0,79	6	8	7
40	2	2	2	1	2	1,75	0,44	4	5	5
41	2	2	2	2	2	2	0,50	6	6	6
42	2	2	2	3	3	2,25	0,83	7	9	8
43	2	2	3	1	3	2	0,79	6	8	7
44	2	2	3	2	3	2,25	0,83	7	9	8
45	2	2	3	3	3	2,5	0,88	8	9	9
46	2	3	1	1	3	1,75	0,76	4	8	6
47	2	3	1	2	3	2	0,79	6	8	7
48	2	3	1	3	3	2,25	0,83	7	9	8
49	2	3	2	1	3	2	0,79	6	8	7
50	2	3	2	2	3	2,25	0,83	7	9	8
51	2	3	2	3	3	2,5	0,88	8	9	9
52	2	3	3	1	3	2,25	0,83	7	9	8
53	2	3	3	2	3	2,5	0,88	8	9	9
54	2	3	3	3	3	2,75	0,94	9	9	9

Fig. 6.8 Vectors (domain) and corresponding scalars (codomain)

					MAX	MEAN	TAU	M 01	M 02	O
55	3	1	1	1	3	1,5	0,73	3	8	6
56	3	1	1	2	3	1,75	0,76	4	8	6
57	3	1	1	3	3	2	0,79	6	8	7
58	3	1	2	1	3	1,75	0,76	4	8	6
59	3	1	2	2	3	2	0,79	6	8	7
60	3	1	2	3	3	2,25	0,83	7	9	8
61	3	1	3	1	3	2	0,79	6	8	7
62	3	1	3	2	3	2,25	0,83	7	9	8
63	3	1	3	3	3	2,5	0,88	8	9	9
64	3	2	1	1	3	1,75	0,76	4	8	6
65	3	2	1	2	3	2	0,79	6	8	7
66	3	2	1	3	3	2,25	0,83	7	9	8
67	3	2	2	1	3	2	0,79	6	8	7
68	3	2	2	2	3	2,25	0,83	7	9	8
69	3	2	2	3	3	2,5	0,88	8	9	9
70	3	2	3	1	3	2,25	0,83	7	9	8
71	3	2	3	2	3	2,5	0,88	8	9	9
72	3	2	3	3	3	2,75	0,94	9	9	9
73	3	3	1	1	3	2	0,79	6	8	7
74	3	3	1	2	3	2,25	0,83	7	9	8
75	3	3	1	3	3	2,5	0,88	8	9	9
76	3	3	2	1	3	2,25	0,83	7	9	8
77	3	3	2	2	3	2,5	0,88	8	9	9
78	3	3	2	3	3	2,75	0,94	9	9	9
79	3	3	3	1	3	2,5	0,88	8	9	9
80	3	3	3	2	3	2,75	0,94	9	9	9
81	3	3	3	3	3	3	1,00	10	10	10

Fig. 6.9 Vectors (domain) and corresponding scalars (codomain)

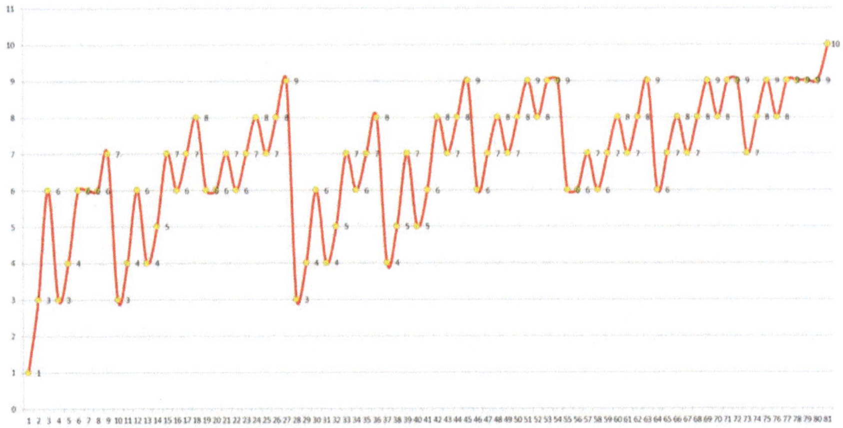

Fig. 6.10 Occurrence diagram

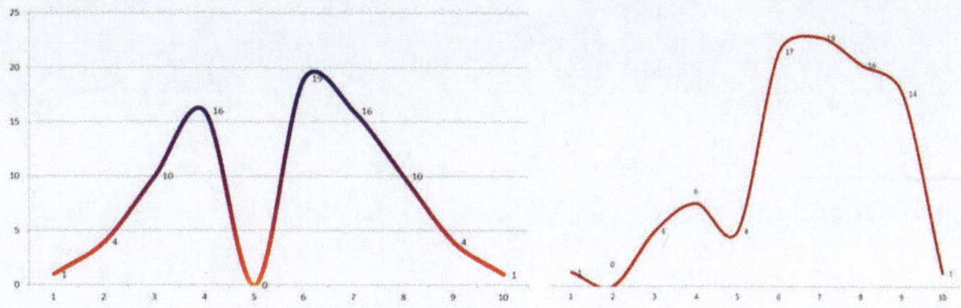

Fig. 6.11 Mean (left) and mean with bias (right)

Final Considerations and Concluding Remarks

7

Abstract

This chapter proposes some considerations on the topics of “Lean Management” and “Lean Thinking”, pointing out hidden dangers in the development of “easier/simplified procedures”.

At the end of this ***conceptual/model-based journey through FMEA***, I would like to add one final personal remark that I consider very important.

Nowadays—and for several decades now—the inspiring idea of “***LEAN***”:

- ***Lean Management***
- ***Lean Thinking***.

and ***similar expressions*** occupy a prominent position within the general framework of quality techniques and methods. These are well known to everyone involved with quality in business at various levels.

We are also well aware that nowadays, “***winning strategies***” are understood to refer exclusively to all action strategies inspired by “flexibility”, “speed of response/adaptation”, “time-to-market speed” and so on.

Now, as long as these are ways of thinking and acting that pursue the systematic goal (absolutely beneficial for any company) of “scientifically” maximizing operational efficiency, I can only agree.

The problem, however, seems to arise when one attempts to ascribe meanings and implications to fundamentally “healthy and physiological” ideas—such as the core concept of “Lean”—that were not present in the original versions and intentions. This attribution is usually quite “unbiased”, as I have observed over many years of work.

Let me give a simple example.

P. L. Belcaro, *9-Object-Model FMEA*, https://doi.org/10.1007/978-3-658-50832-6_7

An activity or process can certainly be "lean"—and that is very good—***provided that this leanness has not been achieved by sacrificing "pieces" of theory and method***, thereby creating ***situations and contexts of potential structural breakdowns***. Specifically regarding FMEA analyses, I would like to caution my kind readers, for example, ***against adopting "simplified/lightened" versions of the FMEA method*** (typically by omitting the ***functional analysis*** entirely or partially).

So far, you have seen an entire "logical edifice" based on the premises and theoretical assumptions of a very precise conceptual model. If one wants to streamline operational aspects underlying the model (i.e., typically procedures), one would first need to thoroughly understand what this would mean ***at the model level***.

Only if the model remains conceptually and logically intact could one then introduce one or perhaps even several streamlining measures without fear. To return to the example: ***conducting FMEAs without functional analysis would necessarily mean irreparably damaging the (vital) hierarchical-systemic relationships***. This carries the risk of losing sight (and understanding …) of the logical connections between the different levels of analysis.

In summary: ***"Lean" is perfectly fine, no problem at all. The only "red alert" I feel I can point out here is that we must ensure this "leanness" is semantically interchangeable with the expression "increased efficiency"***. This increased efficiency must be achieved with scientific rigor, not "by fancy".

Let's always remember:

"ROMA NON FUIT UNA DIE CONDITA"

Rome wasn't built in just a day. I understand that a certain scarcity of resources, time, energy, etc., might necessitate "lean and fast" procedures and tactics. However, let's ask ourselves—always with the utmost care, please—***what rigorous scientific considerations would underlie the simplifications/easements we actually implement***.

I'll close with a joke that I hope will bring a lighthearted and relaxed laugh to all my dear readers. If we intend to "broadcast a simplified program" at the company, perhaps we should make sure it's not something like the news that the Italian comedian Daniele Luttazzi talked about a long time ago:

"This news program will be broadcast in a simplified form to better suit your mental capabilities"

Joking aside, anyone who would like further information, details, clarifications, or anything else regarding anything in this book is welcome to contact me at any time using the following contact details:

MIX
Papier aus verantwortungsvollen Quellen
Paper from responsible sources
FSC® C105338

If you have any concerns about our products,
you can contact us on
ProductSafety@springernature.com

In case Publisher is established outside the EU,
the EU authorized representative is:
Springer Nature Customer Service Center GmbH
Europaplatz 3, 69115 Heidelberg, Germany

Printed by Libri Plureos GmbH
in Hamburg, Germany